W9-ASK-363

DE LA METTRIE'S GHOST

DE LA METTRIE'S GHOST

The story of decisions

Chris Nunn

Macmillan
London New York Melbourne Hong Kong

First published 2005 by
Macmillan
Houndmills, Basingstoke, Hampshire RG21 6XS and
175 Fifth Avenue, New York, N. Y. 10010
Companies and representatives throughout the world

ISBN 1–4039–9495–1

Printed in the U.S.A.

CONTENTS

ACKNOWLEDGEMENTS

I am very grateful indeed to John Bickle and Clare McNiven for their careful criticisms of earlier drafts of this book. They both helped enormously towards eliminating sloppy thinking and phraseology, and in correcting errors of fact. Any remaining mistakes are due to my recalcitrance, despite their best efforts. Many of the ideas described here are less incoherent than they would otherwise have been thanks to input at various times from Harald Atmanspacher, Erhard Bieberich, Julian Candy, Bill Fulford, Stanley Klein, John Sadler and many participants in the Quantum Mind online forum. The task of researching relevant background material was thoroughly enjoyable, due to the skill of the numerous scholars and other writers whose works are named in the References section at the end of the book. I am most indebted to all of them.

Copyright holders who kindly adhered to the 'fair dealing' convention and granted me permission to quote from various works include: Butterworth-Heinemann Publishers; Cambridge University Press; Faber & Faber; HarperCollins Publishers; Oxford University Press; Penguin Books Ltd; Peters, Fraser and Dunlop, agents for the late Professor Norman Cohn; Prentice Hall/Harvester Wheatsheaf Publishers; Routledge Publishers; Servant Publications; and Yale University Press. I would also like to acknowledge permission to include short quotations from Penguin Books, acting for the late Nevill Coghill. I am especially grateful to Professor Benny Shanon for permission to quote extensively from his book in Chapter 13. Some of the material in Chapter 9 has previously been published in the *Journal of Consciousness Studies* and the *Journal of the Royal Society of*

Medicine; thanks to the respective editors of these journals for raising no objection to my using it here. Finally, I was very lucky indeed to get Sara Abdulla as my editor. Many thanks also to her.

Every effort has been made to contact the owners of copyright material which is reproduced here. In the event of a copyright query, please contact the author.

Chapter 1
INTRODUCTION

This book is about who, or what, makes the decisions that we call our own. 'Decision' is a clumsy word, but I couldn't use 'choice' in my title. It would have misled US readers, for it puts many in mind of the abortion debates. I won't be discussing specific choices, but rather what is behind them in general. Is the action wholly due to nerve cells in the brain, or does some other aspect of our nature play a part? Do we really have responsibility for our deeds, or is that feeling an illusion? Are outcomes wholly fixed in advance, or is there some flexibility? All these questions relate to free will or the lack thereof, and getting to grips with them is a bit like being immersed in a complicated detective story. The clues and the evidence needed to uncover what is going on are very varied. Those that I describe range from 21st century discoveries about the chemistry of nerve cells, through the motives behind Brutus's assassination of Julius Caesar, to the curious case of the black pumas. Unlike a detective story, everything that I recount here is based on real events or research findings – with one exception: in two chapters I have woven a tale of a fictional character to show what the clues may mean. Before getting down to it, though, I must set the scene with the ghost of an idea that has, for over two hundred years, haunted our thinking on whether choices can be free.

Julien de la Mettrie, a French doctor caught up in the intellectual ferment of the Enlightenment, was a prolific writer best remembered for just one of his books. More accurately, he is mainly remembered for the title of that book: *L'Homme Machine* or 'Man a Machine'. If matter could be self-organising, as some were beginning to suppose, then life itself might be mechanical and we humans nothing more than complex machines. There would be no need for souls or other mysteries to provide our vital spark. Matter on its own could do all that was necessary to produce a living, feeling human being.

It was an idea that had been incubating for a very long time, boosted by the writings of people like René Descartes and Thomas Hobbes in the previous century. Nevertheless, describing it as clearly as he did proved both de la Mettrie's making and his undoing. He must have known when he wrote that he was pushing his luck with the French Establishment, since he published his work anonymously (in 1747). But his identity was soon uncovered and he had to flee to Prussia. *L'Homme Machine* was banned in France. The implication that his ideas made God redundant was too forceful and too premature. As late as 1766, a far more famous author, Jean-Jacques Rousseau, was expelled from France for atheism. A few years later, though, other French thinkers expressing equally materialistic views remained unhounded for their materialism, if not for their politics. Incidentally, Rousseau found refuge, not in Prussia, but the Peak District of England. He was assisted in doing so by the Scottish philosopher, David Hume, whom we briefly encounter in Chapter 4. Rousseau repaid Hume by developing paranoid delusions about an international conspiracy against himself directed, he thought, by his benefactor. De la Mettrie, on the other hand, kept his sanity and even got a pension from the Prussians, for he was known as something of a *wunderkind*. He died of a fever, still in exile and aged just 43.

The notion that we might be 'nothing but' machines was already implicit in the writings of the Roman poet Lucretius, over 2000 years ago. His opinion that there exist only 'atoms and the void' was taken from Leucippus, Democritus and Epicurus, Greek philosophers of a few centuries previously. Epicurus had tried to soften the consequences of his views by suggesting that atoms sometimes randomly swerve. This might allow the possibility of free will, he thought. Evidently the implications for free will have bothered a few people for a very long time. Throughout most of history, though, such notions have been entertained by a tiny number of maverick thinkers and ignored by almost everyone else. But they have gathered force and permeated popular culture over the last two or three centuries, helped by innumerable writers in the de la Mettrie mould. Most serious scientists in the 17th–19th centuries were theists of one sort or another[1], many also endorsing some concept of soul. In the 20th century this was far from the case. For several generations now, the mechanical metaphor has dominated the thinking of the majority of physiologists and biologists, as well as many psychologists. Feedback between these specialists and the wider public has established the idea as the default setting in many people's thinking.

James Watson, co-discoverer of the double helix structure of DNA, is alleged to have asserted, in a presumably conscious echo of Lucretius: 'there are only atoms... everything else is merely social work'. Watson's former colleague, Francis Crick, became a leading figure in the study of consciousness before his death in July 2004. His popular book, *The Astonishing Hypothesis*, published in 1994, contained the ringing declaration that our joys and sorrows, our memories and ambitions, our sense of identity and free will, are 'no more than the behavior of a vast assembly of nerve cells and their associated molecules'. In one way, Crick's hypothesis is not astonishing at all, because the mechanical metaphor has become so ingrained in our world-view. Yet it is indeed surprising because entirely counter-intuitive to our personal experience. There is an ever-growing industry devoted to tackling what David Chalmers[2] has delineated 'the hard problem', namely how it is that the rich world of our subjective experience could possibly arise from goings-on in a bunch of nerve cells. Success in this endeavour remains elusive.

The problems are especially pressing when it comes to free will. The mechanical metaphor appears to imply that all our actions must ultimately be determined either by physical law or by chance. After all, machines do only what they must. A vacuum cleaner cannot suddenly decide that it would like to be a dishwasher today. A computer has no choice but to follow its programming. Everyday experience, though, hints at something quite different about ourselves. We feel that we are able to choose what we shall do. If pressed, most of us would admit that many of our 'choices' are in fact random or else determined by outside influences. We would nevertheless be reluctant to give up the idea that we have a core ability to freely decide for ourselves what we shall do, independently of any external or internal pressures. Moreover, many of our social structures, including the entire criminal justice system, are built on the premise that people are usually responsible for their actions. It would be worrying, to say the least, to suppose that this apparent capacity for responsibility might not really exist[3].

The almost daily triumphs of modern molecular biology and neuroscience all point to the value of the mechanical metaphor. De la Mettrie was clearly a true prophet, whose approach has born amazing fruit. Regarding bodies and brains as biochemical machinery has allowed discoveries that have doubled our average life span since his time and produced treatments for disease that he could not have imagined. All sorts of advances are rapidly eliminating possible let-outs for those wanting to deny that consciousness is solely a product of our brains. It seems only a matter of time before the last escape route is closed. If our brains are machines, so too must be our minds. That, at least, is how the situation appears to many. Most people in the cultural mainstream are probably prejudiced towards thinking this way.

There are various approaches to squaring the circle – to trying to reconcile the usefulness and predictive power of mechanical models with the fact that they simply don't fit in with how we feel about ourselves. Most of these approaches take free will to be something of an illusion. Many of the philosophers currently writing on the topic, and some of the psychologists, regard the idea as a construct of what they refer to as 'folk psychology', which has little or no validity in their opinion. My approach is different. I propose a more fruitful metaphor for the machine one. We are, I argue, more like stories or films than like machines. Of course stories have to be written with pens or word processors and put into books, while films must be shown through projectors or on television sets. Apparatus of this type has its parallels in ourselves, and the machine metaphor is entirely apt in that context. It is not apt, though, in relation to the essence of the story or film. I intend to show that free will and responsibility are properties of the tale, not of the apparatus with which it is told.

I wish not to exorcise the ghost of de la Mettrie so much as to lay it in its proper context – to show how the spirit of his ideas, the ghost *of* the machine metaphor, fits in with an adequate picture of what we are. There is also another ghost: the one that is *in* the de la Mettrie machine. This second spectre is the essence of ourselves, which has traditionally been termed a soul; I will show how it shares, along with the machine, a degree of responsibility for our actions. It's worth pointing out that, while de la Mettrie's critics were right to suppose that the mechanical metaphor tends to make God redundant, the story metaphor is neutral in this respect. We, along with our relatives, associates, teachers and societies, may be the sole authors and audiences of what we are. Or we may not.

If we are indeed like stories, we are tales that write themselves through the mechanistic apparatus of our brains. In that case, some might object, we must really be machines after all. One of my main aims in what follows, therefore, is to put flesh on the 'story' metaphor and to show that this apparently plausible objection is irrelevant. Starting with some very basic assumptions of the mechanistic view of mind, ones that would have been entirely acceptable to Crick and probably to Lucretius himself, I develop a picture of free will and some other aspects of mind that is wholly compatible with everyday experience. The new picture will give more detail than folk psychology provides, but is no different from a user's own point of view.

Perhaps the most interesting thing about this way of looking at ourselves, at least for those of us who tend to see things from de la Mettrie's viewpoint, is an implication that the really significant constraints on individual free will are not what might be expected. They are not due to physical law or the computational nature of neural information processing. They are in fact due to the story lines available to people. As this is relatively unexplored territory, I examine it in some detail. This picture of a story line shows that it can be thought of as a set of meaningful concepts which are either embodied in people or serve to link them. There is a sense, it turns out, in which the stories connected with a person can sometimes *become* that person. At other times these stories can appear to act according to their own logic, and in ways that people under their influence would not normally have chosen. Indeed, de la Mettrie's proposals were themselves a story line of this type, though the examples I give are mostly of simpler tales.

The notes to each chapter (see p. 193) provide extra information to that in the main text, for anyone who may want a bit more detail. They can be skipped, without risk of missing anything essential. If in doubt, take a look at the three notes to this introduction: they give the flavour of the rest. As to sources, the journal *Nature* has proved invaluable due to its rigour and wide coverage. *The Journal of Consciousness Studies*, too, has provided important material. It has been going since 1994, and was the first multidisciplinary journal to be devoted to the topic. A lot of what will be discussed here has been aired in its pages at one time or another. Other people's ideas are described in the text or can be inferred from the references provided. The overall conclusions however, along with any errors or misconceptions, are my own and cannot be blamed on other writers!

Chapter 2
BACK TO BASICS

Free will is primarily a property of consciousness. It is something that we consciously experience ourselves enacting. So what do modern philosophy and science have to tell us about the nature of conscious experience of any type? An enormous amount has been written and speculated on the topic over the last twenty or so years, so let's take a whistle-stop tour of the main themes – starting with philosophy.

Philosophy

We think we know what consciousness is until we try to describe or define it. St Augustine said much the same about time. Perhaps the most widely accepted brief definition derives from a very influential paper by the philosopher Thomas Nagel entitled 'What is it like to be a bat?' written in 1974, which was one of the chief triggers for the current surge of interest in consciousness. Nagel's definition goes: 'consciousness is that which exists when there is something that it is like to be that thing'. You can pick the definition up in your mind, so to speak, turn it over or look at it back to front, and it still seems quite opaque. But at least, while obviously self-referential in some elusive way, it does not seem entirely circular. The problem with getting at what it means is that one soon gets lost when trying to decide for example what the 'it' in 'it is like' refers to, or whether the 'something' and the 'thing' are the same. But then the subjective essence of conscious awareness *is* hard to put into words, however many one uses.

A more pragmatic definition, used either implicitly or explicitly by a great many people, is: 'consciousness is the difference we perceive when we awake from dreamless sleep'. It's pretty banal, though, and not obviously helpful.

Clearly definitions alone are unlikely to take us far, and we need to look at more detailed discussions. Unfortunately, outsiders may sometimes suspect (rightly) that the opinions arrived at in these arguments are not only confusing but confused, despite the enviable intellectual gifts of many contributors. For instance, it is common for philosophical debates to include what has been called the argument from personal incredulity, i.e. 'I can't believe that, so it must be wrong'. Despite the difficult concepts and occasional dubious argument, it is possible to pick out some areas of agreement and to delineate some of the main schools of thought.

The most important area of agreement has to do with rejecting the belief that Cartesian (substance) dualism is a sensible way of looking at the world. In the 17th century René Descartes divided the world into two fundamentally different 'substances'. There was, he argued, the *res extensa* of our bodies, brains and beyond, which comprised the material world, and the *res cogitans* of our intellectual, conscious selves, which was non-material. The last big name philosopher who wrote about consciousness and argued for a division of this type seems to have been Karl Popper[1]. Almost all now apparently agree that there is only one substance, namely matter or mass/energy, though there are still a few idealists[2] around who suppose that the single substance is not matter but mind. 'Substance dualism' hardly ever gets a serious mention these days by professional philosophers, though some religious and other thinkers still go along with Descartes. Except for the remaining idealists, philosophers appear generally to have settled for materialism, but this comes in a wide variety of flavours.

The toughest-minded philosophers of all, now few in number, are the 'eliminative materialists'. These people say that there is nothing but matter, so it is as pointless to talk about consciousness as it would be to discuss unicorns. We can ignore eliminative materialists, for our purposes, since their stance does not allow them to attach *any* real meaning to the experience of free will. Slightly less austere are the 'identity theorists', perhaps the best known of whom are the San Diego-based husband-and-wife team of Patricia and Paul Churchland. These hold that consciousness simply *is* the brain in action or, if not the whole brain, some particular activity within it yet to be elucidated. To caricature the views of identity theorists only slightly, they argue that there is not much point in talking about consciousness as such. What we should all get on with doing is identifying the brain states which *are* particular conscious experiences. Some of them, especially those writing in the 1980s, looked forward to a day when, instead of saying to ourselves 'What a lovely sunset I'm seeing', we would exclaim 'My brain is in state WX/36/Z' (or whatever classification had been devised). This view has softened a little over the last decade or so, due to the realisation that brains/minds are not isolated objects but are embedded in bodies and environments, which greatly influence their experience. Minds are always 'embodied'[3], to use the philosophy parlance. All the same, the notion that particular brain states equate in some way to particular conscious experiences is at the basis of most current philosophy of mind, even though not all agree with the identity theorists that the brain states simply *are* the experiences.

Daniel Dennett, a prolific and influential writer, looks at these problems from a rather different angle. His 1991 book, *Consciousness Explained*, had quite an impact, as many people understood him to be saying that consciousness is nothing more than an illusion. They were half right about what he meant, though his argument was more subtle than some of them thought. He was much impressed by empirical findings about visual consciousness, particularly those proving that some of what we feel we perceive is illusory. For instance, we are not aware of the quite large blind spot on our retinas that we all have because the visual system fills in the blank area with a sort of averaged version of its immediate surroundings. Our vision appears seamless to us, and proving to ourselves that we really do have a blind spot takes some effort. It involves drawing two small crosses a few centimeters apart on a piece of paper. The next step is to gaze at one cross, through one eye only, while moving the paper closer to one's face. The second dot disappears when the paper is at a particular distance. Findings like this, and there are a lot of them, show that the perceptions we are conscious of are constructed by the brain. They are not some sort of photograph of what is out there. Dennett wondered whether other aspects of consciousness might also be illusory to some extent. There is in fact plenty of evidence that this is so, especially in relation to things like the apparent timing of experience. Although at one time he often did seem close to saying that consciousness as such can be regarded as an illusion, he nowadays appears to go along with the common-sense view that consciousness itself is 'real' in some sense, even though its *content* is a construct that can be regarded as illusory.

Other influential lines of thought grew out of the relationship between information and experience. Clearly there is *some* relationship in that our consciousness contains all sorts of information, much of it about the external world or our own bodies. We know things, but then so does a thermostat – it 'knows' whether a room is hot or cold. So what is the difference between our conscious knowledge and what a thermostat knows? The most radical answer to this question is 'not much'. In this view, aka 'pan-psychism', absolutely everything is conscious in some way. One recent suggestion, for instance, is that the photons permeating all matter are aware[4]. A slightly weaker version of this position is termed 'property dualism' (to be distinguished from Cartesian 'substance dualism'). Property dualists reckon that all matter and/or information has a potential for conscious awareness, which is able to manifest itself in our brains.

'Dual aspect theorists', on the other hand, hold that conscious experience is simply what information looks like from the inside. Normally, from a third person point of view, information is just information. In a system capable of taking a first person, subjective view of information, though, that information will be consciously experienced. Dual aspect theory, when combined with ideas about subjectivity or 'the view from within' as it is sometimes called, is certainly plausible. But it does share a sort of *ad hoc* quality with its cruder cousins, property dualism and pan-psychism. It provides an explanation of consciousness that does not really explain anything. This does not necessarily imply that the idea is useless, since much the same could be said about Newton's concept of gravity. He simply asserted that gravitational action at a distance exists, contrary to all common sense, and this proved a very useful idea. Similarly, the assertion that consciousness *is* potentially an aspect of information is also seen as useful by some. On the other hand, Newton's assertion was backed by mathematics and successful predictions of planetary motions. No analogous successes are visible, not even on the horizon, in consciousness studies. It is rather as if, having abolished Descartes' *res cogitans*, philosophers on the whole could not quite cope with the consequences. They backtracked to some extent and said in effect: 'Hey, we haven't really got rid of it. It was just lurking there in matter all the time. All we've done is get rid of the ontological separation between *cogitans* and *extensa*'. This suggestion does bridge the gap between goings-on in neurons and the vivid world of conscious experience, but it feels just a little too glib.

Does consciousness do anything?

Another theme that preoccupied some philosophers, especially ten or twenty years ago, is absolutely crucial to free will. This is the question of whether or not consciousness is epiphenomenal. An epiphenomenon is something associated with a process which has absolutely no effect on that process. The classic example is the whistle of a steam locomotive, which has no effect on the workings of the engine (except for a tiny power drain). A more up-to-date analogy would be the sound made by a car engine, which is entirely irrelevant as far as powering the car is concerned. Similarly, some have supposed that consciousness is an epiphenomenon of brain function, and can therefore have no effect on it.

It's a question that gets less discussion now than in the 1990s, perhaps because people who said 'yes it is' and 'no it isn't' were about evenly balanced, in numbers and force of argument. Exchanges of the 'I'm right!', 'No you're not!' type quickly make way for more profitable discussions. One of the most cogent contributions to the debates was actually made by a physicist (Avshalom Elitzur), not a philosopher. Elitzur pointed out that consciousness cannot be purely epiphenomenal if only because it causes so much argument about the problems of consciousness! All the same, if people who argued the opposite point of view were right and consciousness is in fact epiphenomenal, any picture of free will anything like the folk psychological version would seem to be completely out of the question.

Identity theorists, who say that consciousness *is* some aspect of brain function, also say that it is not an epiphenomenon. The brain activity that is consciousness must inevitably be capable of affecting its own future functioning and that of the rest of the brain. Pan-psychists, property dualists and dual aspect theorists are in much the same position, since, from their points of view, it is basically a matter of semantics or relative perspective whether one refers to something as a neuronal event or as an experience. One might wonder whether dual aspect theorists have to agree with this, since they need not refer to physical events in neurons. They can confine themselves to discussing the dual aspects of information, which seems a more abstract sort of thing. But the best general definition of information is due to Gregory Bateson, a noted anthropologist and cyberneticist who died in 1980; namely 'information is a difference that makes a difference'. The essence of an epiphenomenon is that it does *not* make a difference. Dual aspect theorists, too, are therefore committed to viewing consciousness as efficacious (i.e. non-epiphenomenal).

People who, like Daniel Dennett, think that consciousness is something a bit like an illusion *can* argue that it is not efficacious. On the other hand, few of them would deny that consciousness is either a product of, or intimately related to, neural activity of some sort – activity which must itself be efficacious. Therefore, to say that *it* is ineffective, while what it describes (the neural activity with which it is associated) cannot be so, amounts to little more than a semantic quibble. For all practical purposes, even if these theorists are right in some abstract sense to distinguish consciousness from its neural entanglements, it must nevertheless be regarded as efficacious. This conclusion about the practicalities holds good unless the distinction between consciousness and brain amounts to some sort of real separability having observable and/or experiential consequences.

There are indeed thinkers, mostly on the fringes of the philosophical mainstream or beyond, who believe that consciousness can be separate from brain in a knowable way. Charles Tart, a US psychologist and parapsychologist, is an especially prominent and sensible advocate of this belief and has written extensively about his reasons for holding it. Some suggest, for example, that phenomena like out-of-body or near-death experiences may be examples of this peeling apart of mind and body. However, if they use experiential evidence of any sort to support their view, they must regard consciousness as efficacious (which they all do, so far as I know). They have to do so because the evidence depends on people's reports of their experience. Such reports can only be made if all sorts of memory-related and other neural functions come into play. Therefore, even if consciousness is genuinely separated from the brain to some extent in the course of a near-death experience, it must still be able to affect the brain. If it had no effect, we would not know about such experiences because we would have no reports.

In short, the only philosophers who can consistently hold that consciousness is an epiphenomenon are committed to a semantic quibble, devoid of any consequences for all practical purposes. It is unfortunate that a good deal of loose talk about how consciousness may not really do anything has seeped into popular culture, helping to bolster the view that free will is a will-o'-the-wisp. There are all sorts of illusory aspects to the experience of free will, just as there are illusory aspects to any conscious experience, but one of the apparently most formidable philosophical arguments against it is groundless.

So the attribution of epiphenomenality to consciousness turns on precisely what is meant by the term 'consciousness'. The argument can be made only if consciousness can be distinguished in some meaningful way from all neural activity of any sort. Rejection of substance dualism already makes any such claim distinctly dubious. What finally kills it is that, if the distinction were sufficiently real to allow experiential or other detection, then the argument must be untrue because detection requires that the consciousness in question be reportable in some way. This in turn means that it must have effects on the brain, even if only on memory systems and, via them, on verbal or other motor systems. In fact, as we shall see, it is precisely on memory systems that consciousness has its main impact.

What has philosophy told us?

Contemporary philosophers make many interesting points about consciousness, especially in connection with how it relates to the body it inhabits and the subtleties associated with subjectivity. Practically all agree that our form of consciousness is associated with brain activity. They disagree, often passionately, on whether it can be identified with brain activity, whether it may be some more general phenomenon that the brain taps into, whether the brain generates it in some way, and so on. Nevertheless, if pressed, nearly all accept that, if you have some particular and unique conscious experience, then there will exist some particular and unique neural state or activity that either *is* the conscious experience, or generates it, or at the very least closely correlates with it. Note that the reverse is not true. Only a small subset of unique neural states are associated with unique conscious states, since most neural activity is *un*conscious. It takes uniqueness of neural activity plus some other, unknown, ingredient to either *be*, or be intimately linked with, a unique conscious state. The influential modern philosopher Jaegwon Kim made the same point when he said: 'Any two things that are exact physical duplicates are exact psychological duplicates as well'. The 'psychological duplication' might refer, among other things, to absence of conscious awareness or its presence.

This rather general point provides one of two main foundation stones (axioms) needed for building an adequate concept of free choice. Here it is again, more succinctly:

> Any distinctive conscious experience is associated with a distinctive neural state or activity.

This statement, though derived from current philosophical opinion, is actually the basis of most present-day neuroscientific research on consciousness. It is often implicit, but Francis Crick in particular made it explicit. The research programme that he advocated involves looking for the neural activity that is most exclusively correlated with particular visual experiences. It is a programme that still has a long way to go. We can access distinctive experiences simply through introspection – seeing a red wall is a different experience from seeing a blue one, for instance, and people can report this to investigators – but we don't know enough, or have sufficiently sensitive techniques, to distinguish the two neural states that we assume are directly associated with the two experiences, even though it would probably be just about possible with present-day techniques to tell whether someone was experiencing red or blue by examining their brain activity. Certainly it would be quite easy to tell whether they were seeing a colour or hearing a sound. But the differences we can detect at present may, for all we know, refer to *un*conscious aspects of perception, not conscious ones. We don't yet know, in other words, whether the neural activities that we are able to detect are those most closely correlated with the conscious experience itself. So what do (some) scientists think about consciousness?

Neuroscience

Neuroscientific theories are even more varied, especially in their detail, than philosophical ones. It is often not clear how much different theories overlap, or even whether some may or may not be mutually exclusive. But common themes stand out. Perhaps the largest theme, a sort of Amazon among the various rivers of thought, is that consciousness has to do with recurrent, self-referential nerve activity.

Douglas Hofstadter gave the self-referential part of the package a huge boost in his 1979 book, *Gödel, Escher, Bach: An Eternal Golden Braid*. The book caught the imaginations of numerous people, including many of the brain scientists who were then just beginning to think about what sort of thing consciousness could be. In it, Hofstadter describes the wonders of J. S. Bach's music, M. C. Escher's lithographs and Kurt Gödel's mathematical proof that there are true arithmetical theorems that can never be proved arithmetically. All these seemingly very disparate achievements crucially double back on themselves, so to speak: one part refers to the whole and vice versa. Using a very nice parable about a 'conscious' ant colony, Hofstadter implies that consciousness itself relies on similar self-reference. There are two main suggestions as to how this self-reference could occur in the brain.

The suggestion that attracted most attention at first has to do with recurrent loops of nervous activity between sub-cortical centres in the brain and the cerebral hemispheres, called cortico-thalamic loops. Erich Harth's 1993 book *The Creative Loop: How the Brain Makes a Mind* gives an especially readable account of this type of model. Anatomists have certainly found these loops and there is evidence that they are important to functions closely associated with consciousness, particularly attention itself. Recently, however, many researchers have relegated such loops to a supporting role. The discovery that, when separate brain areas are dealing with a single percept, such as looking at a friend's face, their electrical activity is often synchronised[5] led many neuroscientists to suspect that intra-cortical circuitry might be the basis for consciousness. Cortico-thalamic loops may help other feedback loops to trigger cortical synchronisations. Others, notably the Nobel Prize winner Gerald Edelman, hold that ordinary recurrence is not strong enough to provide the sort of unified basis that consciousness needs. They argue that 're-entrant' nerve activity is required. This involves sets of feedback circuits, in which the information in one set is constantly and reciprocally updated from the information in others.

Another idea that has a lot of evidence going for it is that consciousness is a function of a 'global workspace'. This is the view that San Diego-based psychologist Bernard Baars, in particular, advocates. The picture here is that lots of specialised, but unconscious, mental modules process information. They constantly compete for access to the hub of a centralised distribution network. The winning modules get to have their information distributed to all the others. This ever-changing, widely shared information *is* the stream of consciousness in this model.

Many neuroscientists picture consciousness as being divisible into a sort of background, general capacity for conscious awareness, and a variable number of more specific facilities. Self-awareness is always among these. The conscious 'I' tends to be viewed as an amalgam of neural self-models in the brain with aspects of its emotional and stabilising functions. Some single out language as being fundamental to a specifically human capacity for consciousness.

Today's most popular de la Mettrie concept of the brain likens it to a computer. Descartes himself pictured the brain as an hydraulic machine while, early in the 20th century, people often compared it to a telephone exchange or endorsed Sir Charles Sherrington's famous analogy with an 'enchanted loom'. There is a bit less enthusiasm for the computer analogy now than there was ten or twenty years ago, mainly because of the embarrassing discovery that Artificial Intelligence is hard to achieve. Brains can easily do things that are difficult or impossible for computers. Despite recent leaps in technology, recognising individual faces from different angles and in changing light conditions, for instance, is still hard for computers but easy for us. In fairness, even we have trouble recognising faces upside down. The brain, neuroscience now concedes, cannot be very like any of our present-day computers. Nevertheless there must be something in the analogy since both brains and computers process information, while consciousness contains an ever-changing stream of information about all sorts of things, not unlike the data on the screen of a working laptop. There is, therefore, a lot of interest in how the brain encodes information, particularly conscious information.

Nerve cells transmit information to one another by electrical and chemical activity. In many respects the electricity and the chemistry are two sides of the same coin: electrical changes arise from, and influence the movement of, chemical ions. Christof Koch, who worked with Francis Crick for many years, described in 1999 how there are 14 separate mechanisms within individual neurons that could undertake computation. It's not known how many of these mechanisms are actually important. Nevertheless, individual brain cells are certainly far more sophisticated than the transistors in a computer.

There is another layer of computation over whatever is confined to individual cells. This relates to communication networks between cells. The first transmission mechanism to be discovered, still often viewed as the most important, involves action potentials, all-or-nothing electrical waves that sweep along nerve fibres. These can trigger the release of more than fifty neurotransmitter or neuromodulator chemicals, which may in turn cause an action potential in another neuron or modify its probability of firing. Information is certainly coded in the frequency of action potentials and is probably often also coded in their timing, independently of overall frequency.

Neurons communicate in other ways also. For example, some cells release hormone-like substances into the bloodstream which affect other parts of the brain. Local diffusion of various chemicals can play a part as well. Waves of varying calcium concentration spreading through support cells can affect nearby nerve cells. Nitric oxide, a gas which can diffuse quickly through the brain over short distances, may be important too[6]. Local electrical fields are probably significant, particularly in affecting the finest nerve endings, called dendrites. Indeed, although cell membranes normally insulate the innards of nerve cells from one another, there are direct electrical connections, called gap junctions, in some places that aid the spread of electrical fields. The densest concentration of these is between dendrites belonging to different neurons.

All these mechanisms relate to how the brain may deal with information. Their connection, if any, with consciousness, is unclear. As we will see shortly, there is only indirect evidence about how the information in consciousness itself might be encoded. It seems unlikely that conscious experience could be directly based on things like the frequency with which some particular neuron fires, since it normally relates to higher level, more integrated, brain functions. That is not to say that the firing of an individual neuron can never register in consciousness, since this probably can happen sometimes. For instance, recordings taken from parts of the brain that recognise particular faces or places show that very few neurons seem to be responsible – and we are of course conscious of things of this sort. Moreover direct electrical stimulation of just a few cells can sometimes produce a conscious experience. All the same, an individual cell can be expected to register by virtue of affecting some extensive property or activity involving many neurons. Few people hold that an isolated nerve cell could be conscious.

Much of what happens in the brain seems to be poised on the borderline between order and chaos. Edge-of-chaos systems have a fractal structure and are widespread in the brain at all sorts of functional levels, from the anatomy of dendrites, through the electrical activity of whole brains, to some aspects of the organisation of cognitions. Systems like this are often very sensitive to small inputs: activity in a single nerve cell can sometimes trigger a big change in the state of the whole system. But it is the state of the whole, or at least of some part of it much larger than the single neuron, that is likely to relate most directly to consciousness.

So it is understandable, against this background, that Walter Freeman and his colleagues, working in California in the 1980s, found evidence that what rabbits are conscious of smelling relates to the pattern of electrical chaos in their olfactory lobes. Freeman gave his rabbits something to smell while measuring the electrical field changes in the parts of their brains that process information about odours. Interestingly enough, the same smell (meaning the same odorant applied to the rabbit's nose) never gave rise to exactly the same type of chaotic activity twice, though there were more similarities in the activity with the same smell than with a completely different smell. This suggested that rabbits' consciousness might be encoded in chaotic activity in some fairly direct manner. After all, we ourselves do not have identical experiences in response to identical stimuli – think of the difference between smelling new-baked bread when you are hungry and when you have eaten too much – and maybe rabbits are like us in this respect.

Quantum mind theories

A completely different line of scientific speculation about consciousness is that it might have some quantum theoretical basis. Physicists signed up to the traditional Copenhagen interpretation of quantum mechanics, needed to understand what conscious observers really are, since these play a central part in their theories about matter. Conscious observers 'collapse the wave-function' in the Copenhagen view, resulting in an object being observed in just one of its possible states; they are thus responsible for either killing or reprieving Schrödinger's famous cat. Other theorists[7] felt that quantum field theory might provide a solution to the binding problem: how disparate activities in the brain are unified into a single conscious experience. Yet others had the vague feeling that, since both quantum theory and consciousness are great mysteries, they must be connected.

Interest in quantum mind theories peaked in the early 1990s, probably due mainly to Roger Penrose's book *The Emperor's New Mind*, published in 1989. This is a fascinating popular account of fundamental physics and its possible connections with mind, by a very eminent mathematical physicist who, among many other achievements, worked with Stephen Hawking on the theory of black holes. Today the quantum theories of mind are no longer the hot topic that they were. More physically plausible solutions to the binding problem have been suggested. People are now less interested in the Copenhagen interpretation of quantum mechanics. Conscious observers have largely lost their central role in fundamental physics to an automatic process called decoherence. Furthermore, the growing body of research on quantum computation has made people think it very unlikely the brain could use any such mechanism. It seems far too fragile to survive such a warm, wet, messy environment.

None of these objections absolutely rules quantum theory out of consciousness. That it might have a role is still being actively investigated. One of the main theories[8] involves quantum effects acting in structures inside neurons called microtubules. Other suggestions are that fields of exotic quantum particles of various types may affect brain function[9]. Many of these ideas are highly technical and difficult to follow for most neuroscientists, untrained in quantum field theory. It is fair to say that there is no convincing empirical evidence in favour of any of the quantum theories of mind. So far.

What neuroscientists do

Mostly neuroscientists who are studying human consciousness ask people to report on what they are experiencing, or have experienced, in various experimental settings. But people can only report on a conscious experience if they can remember it. Neuroscientists study *reportable* consciousness, which must be closely related to memory of some sort.

The same limitation exists for all of us, even when we are just mulling privately over our own conscious experience. If we had a good meal yesterday, or a month ago, we may remember the taste of the main course, or we may have recorded what it was like in a diary, or in some other way. But if there is no memory or other record, we have absolutely no way of knowing whether or not we were conscious of the taste of the food at all, or indeed of whether the meal ever happened. The crucial point is: exactly the same applies for an experience that we had half a second ago. Memory, here, over the half-second period, provides the only possible record of experience. We cannot possibly be aware of any supposedly conscious experience unless it gets remembered, at least for long enough for us to introspect it. In other words, even our private conscious worlds deal only with the reportable (at least to ourselves). Anything that does not get remembered for long enough to be introspectable is, for all practical purposes, unconscious.

Doesn't one have to be able to introspect the introspection that one had introspected? The simplest way to avoid this brain-teaser is to bite the bullet and say that consciousness *is* a memory-related function, at least as far as our personal experience and any neuroscientific investigation are concerned. Philosophers and others may wish to discuss some concept of non-reportable consciousness, but it is absolutely impossible to prove whether any such entity exists. It is far more reasonable to take the view that non-reportable 'consciousness' is not consciousness at all. 'Knowledgeability' might be a satisfactory term for it. We could then say that thermostats 'know' whether a room is too hot, but are not conscious of temperature because they don't know that they know how hot it is. If people did not remember their own conscious experiences, they certainly could not tell investigators about them, nor could they tell themselves, so there would be no possibility whatsoever of discussing problems of consciousness. They would, to all intents and purposes, be forever unconscious, like perpetual non-dreaming sleepwalkers. It is memory that saves us from this fate.

There are potential pitfalls to using the term *reportable*. A typical consciousness experiment, for example, might take the form of flashing emotionally loaded pictures (of people fighting or kissing, say) onto a screen so briefly that they are not seen. To achieve invisibility to conscious perception, it is usually necessary to follow them immediately with some neutral, 'masking' picture, such as a photograph of a tree or a beach scene. Then the researcher gradually increases the time for which the emotion-producing pictures are shown until the subject reports seeing them – verbally or by pressing a button, for instance. Typically, some measure of emotional arousal, such as pupil dilation, occurs in response to briefer presentation times than those that elicit a verbal or press-button report. Isn't pupil dilation a 'report' of a sort? Indeed it is, but it is not a report of a *conscious* experience, it is only an indication that the brain has registered information about something. It is regarded as a measure of *unconscious* perception. The only responses which are regarded as reports of conscious experience are those in which subjects explicitly or implicitly state 'I recall my consciousness of this experience'. Most psychologists agree that all sorts of unconscious brain activities occur – perceptions, cognitions, even emotions – which differ from their conscious counterparts.

Surely animals cannot report their conscious experience, so how can they be used in consciousness studies? Certainly few Behaviourists, who at one time dominated academic psychology, allowed that animals might possess consciousness. But the pendulum has swung and most people now agree that it is in fact reasonable to *infer* that they are often conscious, even though we cannot ask them directly. In the case of some animals the grounds for the inference are strong. These creatures have vocabularies. They can refer to themselves in the 'first person', describe their likes and dislikes, and even their hopes and fears. Alex the parrot, for instance, the subject of research by Irene Pepperberg of the Alex Foundation, asked one day to be taken for a ride in the car and expressed considerable annoyance when told the car wasn't available! Sue Savage-Rumbaugh of Georgia State University, among others, has studied vocabulary and self-awareness in apes, especially bonobos (the chimp-like creatures that prefer making love to making war, and which are being exterminated by hungry humans). She concludes that they are at least as conscious as pre-school children.

It seems entirely reasonable to think that these animals have consciousness, reportable to themselves, that we can tap into in experiments. One type of study on monkeys, for example, involves getting them to look at ambiguous drawings (the ones that can be seen as either a duck or a rabbit, as a man's face or a seated girl, etc.), and communicate which interpretation they experience; meanwhile the electrical activity in their brains is recorded in a variety of ways. The percepts differ, but the information reaching their brains is always the same, so any differences in brain activity must reflect their conscious experience, not the input. This is clever technology which is producing lots of clues about consciousness but no definite answers yet.

Y

That scientific study of consciousness must always refer to *reportable* consciousness has been regarded as almost embarrassing by some people, as if it implies that science cannot get to grips with 'real' consciousness. Bernard Baars, of 'global workspace' fame, and his colleague Katharine McGovern, for instance, forcefully pointed out that science can deal only with reports of experience, and then went on to imply that this is some sort of limitation *despite which* we must simply get on and do the best we can. In actual fact, if it is a limitation, it is one that helps to make sense of the whole field of study, and is especially important to developing an account of free will that has both scientific and intuitive validity.

There is a lot of evidence (more of which later) that consciousness of information reaching the brain takes time to develop; around one fifth to as much as half a second depending on circumstances. Clearly it *must* in some sense be memory-related[10]. Similarly, the view that consciousness is memory-related is entirely consistent with the mainstream belief that it may arise from iterating brain activity[11]. After all, later activity in the circuitry provides a memory of earlier activity.

We can now formulate the second axiom of free will:

Consciousness is a memory-related phenomenon.

This and the first axiom (p. 15) are fully consistent with materialism. There is nothing in them to which the most hard-headed of rational thinkers could object. They are also consistent with current neuroscientific findings and with mainstream theory. Next up: memory.

Chapter 3
MEMORY

Today our knowledge of memory is a bit like a road atlas with a lot of the pages missing. It may give a good description of Devon, say, or the Lake District, but is little help on how to get from one to the other. We know quite a lot about the individual sorts of memory that I'll be describing, but the overall picture, especially of how one type relates to another, is hazy at best. What we ordinarily call memory, psychologists term declarative memory. It is the faculty that enables people to say 'I went to the supermarket yesterday and bought some pizzas and ice-cream'. Important, but mostly unanswered, questions about declarative memory centre on how it is recorded, what form the record takes and how it is recalled. Another form of long-term memory is usually called learning or procedural memory. Once you have learned to ride a bicycle, for example, you may not be able consciously to recall the details of how to do it, or even remember much about your practice sessions. Nevertheless, you don't forget how to ride. In addition to these two long-term forms, there is a range of short and very short-term memories.

What we lack are good maps showing how these short-term varieties relate to each other and to declarative memory and learning. But it's not all bad news. John Bickle, for instance, a Cincinnati-based philosopher who is interested in neuroscience, points out that it is remarkable how much we *do* know about memory. His 2003 book *Philosophy and Neuroscience: a Ruthlessly Reductive Account* provides a particularly engaging round up of recent research on memory-associated neurochemistry and of what is known about the relationship between activity in single neurons and consciousness. A great deal is known or hypothesised about the molecular detail, especially relating to how short-term memories may get consolidated into a long-term form. Very little is known, on the other hand, about the organising principles involved. (For example, exactly how is information selected for translation? Precisely what form or forms does the long-term log take?)

Many of the early insights into the mind and brain came from neurologists studying people with brain injuries. For instance, soldiers with bullet wounds to the backs of their heads from the Russo-Japanese War of 1904–05 and the First World War provided a lot of information about what the visual cortex does. Stroke victims were another mine of information, though neurologists of pre-brain scan days had to wait until their patients had died in order to relate the clinical findings to the areas of brain damage discovered at autopsy. Nowadays scans allow much faster feedback. The famous 19th century patient Phineas Gage, a railway worker who had an accident in which an iron bar went through his temple, damaging his brain, provided insights into the role of the frontal lobes in higher mental functions. Before the accident, Gage was a model citizen. Afterwards he became an idle, ill-tempered layabout, despite an amazingly complete physical recovery from his horrific injury. Sadly, a lot of information about what the frontal lobes do was also gathered from ill effects of the many mid-20th century lobotomy operations carried out in (mostly misguided) attempts to cure psychosis. Parts of these lobes are now known to specialise in different emotional, cognitive, attentional and memory-related functions.

With long-term declarative memory, the evidence from patients suggests that it is a diffuse function, spread across the whole brain. Apart from one exception, no local injury can knock memory out in the same way that losing your visual cortex can render you blind. Rather, it seems that the more extensive any damage, the more degraded memory becomes. It never entirely disappears, but fades away like the Cheshire Cat. The exception is given the name 'Korsakov's psychosis'[1]. It does not completely destroy memory, but punches an enormous hole in it. People who develop the condition show normal declarative memory for events up to the time that they get it. They also show normal memory for everything in the most recent two or three minutes. It is just that there is no ability to recall real events that happened in between. If their illness began when they were aged thirty, and they are now actually fifty, what really occurred in the intervening twenty years simply has not happened as far as they are concerned. Learning (i.e. 'procedural memory'), however, tends to be less affected than declarative memory.

Interestingly, many people with Korsakov's fill in the gaps in their memories. Asked what they did yesterday, they might come up with some story about how they visited the seaside with their families, when in fact they were in a hospital ward all the time. They seem to believe these stories; there is no intention to deceive. A proportion do not even tell stories believable to others, but describe James Bond-type adventures, apparently without any insight that their tales could not possibly be true. The technical term for this filling in of the gaps is 'confabulation', or 'fantastic confabulation' in the case of the James Bond stories. It can occur not only in Korsakov patients but in others – most commonly people with Alzheimer's Disease, which causes neuron death in many brain areas.

The hippocampus

The conclusion that 20th century neurologists drew from these cases was that short-term memories, lasting a few minutes at most, have to be transcribed somehow into permanent memory if they are to be retained. Korsakov patients lose the machinery needed for this. When their brains are autopsied, damage is found to the limbic system, renowned for its emotional functions as well as its role in memory. This system is a large and specialised version of the cortico-thalamic loops, which some neuroscientists believe may provide the physical basis for consciousness. The limbic system links the thalamus and other structures with areas of the cortex. The closest connection is with the oldest bit of the cortex in evolutionary terms (named the cingulate gyrus). A particularly specialised part of this circuitry is the hippocampus; it is this that has attracted the most research interest in recent times.

In Korsakov patients, damage was often found to parts of the limbic system called the mamillary bodies, or sometimes to areas of the thalamus itself. However, in the late 1950s, the hippocampus was revealed to be at the heart of the memory system[2] when a particularly famous patient, known as 'H.M.', developed a Korsakov state after surgery to remove parts of his brain, including both hippocampi. The operation was done in an attempt to cure his very severe epilepsy. The case attracted a great deal of attention from neurologists and psychologists interested in memory, and they were still studying H.M. twenty years later. Further experience with the operation showed that removing only one side of the hippocampus did not greatly affect memory, and that the crucial damage responsible for Korsakov states was to the hippocampus, not to the other parts of the brain that had been removed.

Meanwhile laboratory studies, mostly on rats, were suggesting a rather different picture. This research showed pretty clearly that a main function of the hippocampus has to do with spatial information and memory. Nerve cells were found in it, for example, that fired only when a rat reached some particular part of a familiar maze. How can this be reconciled with the evidence from people that the hippocampus mainly transcribes short-term into declarative memory? There is in fact some data (mostly from animals, but there's no reason to suppose people are any different) suggesting that it stores a wide range of episodic memories (e.g. 'I remember this particular turning in this maze') and links them into the sorts of story that we access in declarative memory (e.g. 'I was running through this maze feeling breathless when I came to a corner I recognised and some idiot blew a whistle in my ear!'). Nerve cells in the hippocampus have an interleaving topography, which appears ideal for linking separate episodic memories. Perhaps the hippocampus's principal function is to organise and coordinate individual short-term memories into a narrative form better fitted to long-term storage.

Working memory

A slightly different perspective on the limbic system in general, and the hippocampus in particular, is to think of it as an important centre for 'working' memory, though the prefrontal cortex is also involved. Other specialised areas seem to get drawn in, too, depending on what is being remembered. Working memory holds what is available to awareness at any particular moment. These 'moments' are quite flexible things, ranging from less than a second to a minute or more, depending on circumstances. A figure of around 20 seconds has sometimes been quoted for how long working memory itself holds onto things, but moments of experience are usually a lot briefer. They're only a fraction of a second if you are driving happily along, for instance, and suddenly realise you've hit a patch of black ice. And they can be longer than 20 seconds if you are drowsing comfortably in a chair listening to favourite music. Working memory is thus an important type of short-term memory. It is often pictured as analogous to the RAM in a computer, though the analogy breaks down if expanded to equating declarative memory with ROM on the hard disk. Declarative memory is probably not much like any existing ROM, if only because it is not Read Only (more of which later).

Probably the single most surprising thing about working memory is its small capacity for separate items. We've known, presumably since the dawn of humanity, that we can hold only about seven items in this type of memory at a time[3]. For instance, told a new telephone number, you are unlikely to be able to dial it from immediate recall if it contains more than seven digits. Moreover, it is harder to remember larger items in the longer term. Like many, I can still remember the phone number of my childhood home ('Gosforth 237') but have completely forgotten the 10-digit phone numbers of houses I've occupied more recently, except for the current one. The earlier number is easier to recall because my memory treats 'Gosforth' as a single item, while each separate digit is also an 'item' for most of us (people able to perform memory feats group many digits into 'single items'). There is evidence that our short-term visual memory cannot even manage seven items, but struggles to reach four[4]. This is not so much of a limitation as it might seem, since the information capacity of working memory may still be large. If the separate items are digits, like a phone number, the information held in memory may be as little as 49 bits (the '1's and '0's of computer memory). A single decimal number can be uniquely identified using four bits; its order in a seven-digit sequence can be specified with another three. But the items might be things like books or symphonies. Asked, 'Which is your favourite book?' you would probably be able to compare memories of seven at any one time. In these circumstances the amount of information held must be large[5].

What underpins long-term memory?

Studies of the limbic system have provided one of the two principal themes in memory research. The other stems from a suggestion made by the giant of neuropsychology, Donald Hebb, in the 1940s. He theorised that, if connections between nerve cells that are active at the same time become stronger in some way as a result of their co-activity, this could provide a basis for memory – dubbed 'Hebbian learning'. The strengthening is now usually pictured in terms of an increased probability that an action potential, arriving at a synapse (the junction where one neuron contacts another), will stimulate the next cell to fire[6]. Regular synaptic transmission that triggers an action potential in the next cell may increase this probability. 'Hebbian learning' remained a theoretical possibility for many years, but is now known to occur. Its precise role and importance are still controversial, however[7].

The main messenger chemical in the brain, glutamate, is detected by various receptors on nerve cells. One particular type, the NMDA receptor, shows exactly the sort of changes Hebb predicted. The more that synapses incorporating these receptors are used, the more effect they subsequently have on the membrane potential of the cell that is being stimulated (i.e. the more these synapses are used, the more likely they are, next time they are activated, to cause the stimulated cell to fire off an action potential). The effect has been named 'long-term potentiation' because it can last indefinitely. Interestingly, hippocampal neurons may use similar changes in synaptic efficacy to convert information about the timing of neuron activity into spatial patterns. If so, Hebbian learning is probably important to short-term, as well as long-term, functioning.

We don't know much about how short-term memories get translated into long-term ones, even though the complex biochemistry underpinning long-term potentiation in NMDA receptors *has* been explored in considerable detail. The process must involve some editing, as not everything in short-term memory gets transcribed. It is quite easy to prevent transcription of particular memories by distracting people's attention at the right moment, for instance. The most important piece of information of all to do with this topic is that the process of forming a long-term memory often, or perhaps always, involves protein synthesis. Animals don't retain training if immediately afterwards they are given a substance that inhibits protein synthesis. But such inhibitors don't have any effect on memories once they are in storage. In other words, inhibitors are a bit like Korsakov states in that they prevent short-term memory being transcribed into long-term memory. No doubt the detailed mechanism is different; indeed it must be, because Korsakov patients *can* retain new procedural, though not declarative, memories. For instance, a Korsakov patient can learn new skills, even quite complicated ones such as typing, without being able to consciously recall anything whatsoever about the training sessions they were given. But animals given protein synthesis inhibitors at the right time can't learn skills. This finding about protein synthesis led to the fascinating discovery that long-term memories, too, can probably be edited.

Many of us tend to think of our declarative memories as being a bit like photographs that can be taken out of some sort of mental album and looked at. This is a very misleading analogy because what the brain seems to do, when recalling a long-term memory, is in some sense to re-create the original experience. It doesn't just open a file where the whole thing is stored ready-made. Admittedly the re-creation is usually a rather pale shadow of the original, though not always. Some flashbacks, for instance, are at least as vivid as original experiences, if not more so. Anyhow, it has been found that, if you get an animal to recall a long-term memory and then give it a protein synthesis inhibitor immediately afterwards, the long-term memory subsequently disappears or is greatly weakened. The most plausible explanation is that the memory is returned to short-term, working memory on recall, and must then be re-transcribed (the process requiring protein synthesis) back into a long-term form if it is to be retained.

So it looks as though long-term memories can be edited, perhaps when they are recalled, perhaps when they are re-transcribed after recall, or maybe both. This possibility is very relevant to both psychotherapy and false memory syndrome because it means that a person's memories can be altered, presumably in conformity with their current preoccupations, after having been recalled. Psychotherapy of course involves a lot of delving into people's distant pasts, as do many police enquiries, and it is likely that the memories change – maybe only a little, but possibly quite a lot – each time they are examined. Although false memory syndrome is the downside of this changeability, there is an upside (more of which in Chapter 12): namely that the ability to edit helps free us from the toils of social determinism. It's essential to our free will.

Cellular memories

A lot of memory mechanisms operate at the cellular level, so really serious pruning or editing must be involved in selecting which memories shall get into working memory.

All information reaching the brain probably triggers an incipient memory of some sort at this basic level. Only a tiny proportion ever makes it through to the next stages. For example, the optic nerves transmit data at around seven million bits per second to the visual centres of the brain. Working memory can handle new visual information at a rate of only a few hundred bits per second. The rest, apparently, is sooner or later lost.

Not much is known for sure about the relative importance of the various cellular memory mechanisms. Let us just consider a particularly neat example. Local increases in calcium ion concentrations within nerve cells perform all sorts of signalling functions and are closely related to the amount of electrical activity going on. There is a protein, called CaMKII, widely distributed within neurons, which exists in either an active or an inactive form. The active form has a range of effects, including influencing the wiring of developing brains. It also influences synaptic efficiency and helps with long-term potentiation, thus contributing to Hebbian learning. The really neat thing is that switching to the active form depends on local calcium ion concentration, and the time for which the protein stays switched increases with increasing concentration, until a threshold is reached at which it stays permanently on. In this case, the editing of information at a cellular level is a function of how much the information in question is reflected in increased calcium concentration. Only those (presumably rare) concentrations high enough to switch on CaMKII permanently get remembered for any great length of time. So only the information associated with high concentrations is remembered for long.

Like all proteins in the brain, CaMKII does not last for ever. Its 'half life' is probably about a month, which means that some molecules will last only a few minutes and others for years, but, on average, half the molecules will be gone after a month. But there is evidence that replacement molecules take the same form as the originals, either active or inactive. Memories of the occurrence of high calcium concentration within cells may therefore last indefinitely. The possible functional significance of this wonderful mechanism is purely a matter for speculation at present[8]. There's a strong temptation to think that long-term, declarative memory must be linked in some way with the permanent change in CaMKII that occurs if calcium levels rise high enough, but there could be no connection between the two.

Attention and memory

Attention and memory are intertwined in all sorts of ways. At the most basic level, one cannot sustain attention to anything unless there is some form of memory of what one's attention was doing a moment ago and of what the 'anything' is. Attention is thus entirely dependent for its very moment-to-moment existence on intrinsic memory(ies). But what gets into working memory, and thus ultimately into long-term storage, depends mainly on attention: as every schoolchild tires of hearing, 'If you don't pay attention, you won't learn'. Actually the memories associated with quite short-term attention can last for a surprisingly long time, it seems. If, in an experiment, someone is asked to pay attention to a particular colour (from a choice of four colours) for a brief period, this has effects on visual performance subsequently which last for up to a month. There's a certain amount of debate about whether attention is always needed, particularly for what has been termed 'implicit learning', but there's no doubt that most of what we learn and remember gets into our minds because we've paid attention to it at some stage.

Then again, the objects on which attention focuses seem to be available because they have been remembered. They are, in a sense, objects extracted from memory that happen to coincide with features of the world 'out there'. For instance, the targets of visual attention seem to be pre-learned features of the visual environment, rather than some raw flux of visual input. Masao Ito, a leading vision researcher, put it this way in an editorial in *Nature* in 2000: 'How can we instantly recognise [and thus pay attention to] a familiar object? Probably because, in the brain, we already have a model [i.e. memory] of that object which is activated through vision'. We not only hold memories of objects in the external world. We also hold memories of concepts and ideas – 'objects' of cognition, on which there will be plenty later on.

Clearly attention and memory are entangled over all sorts of timescales. Moment-to-moment attention depends on very short-term memory. But attention then helps to select what gets into working memory, the contents of which influence the future direction of attention. Finally, transcription from working memory into long-term storage provides the 'objects' on which attention can focus. Feedbacks between the two operate over times ranging from less than a second to many decades.

Consciousness and memory

We can now ask: 'To what aspect of memory is consciousness most closely related?'. Cellular memory won't do, since most neuronal goings-on at that level are unconscious. Working memory is a lot more promising as its contents usually *are* conscious. What about the in-between stages, the transcription from cellular to working memory? Practically nothing is known about the processes involved other than that they often or always depend on attention.

In the later stages of the memory process – transcription from short-term or working memory into permanent storage – everything fades back into unconsciousness again. As we saw, conscious recall of a declarative memory involves its coming back into working memory. The most reasonable conclusion to draw, therefore, is that consciousness has some intimate relationship with working memory, perhaps especially with how this is extracted from cellular memories. Consciousness, in brief, is at the very point at which the mass of information, both that reaching the brain from outside and what is within, gets pruned, edited, sorted and sent (or re-sent) for long-term storage.

Although it is not possible to say anything definite at present about *how* consciousness might contribute to editing all this information, the mere fact that it must be involved in the process *somehow* is enough to let us build a picture of how free will works. Placement at this crucial point in the memory process where selection, pruning and filing takes place is what allows our conscious selves a degree of responsibility and free choice in relation to our brains. Let's find out why.

Chapter 4

FREE WILL AND FREE WON'T

Things have been looking particularly black for free will since the 1960s. De la Mettrie and similar thinkers already had difficulty finding a place for it in a universe based on Newtonian-type law and chance. Darwinism increased the problems. Then the computer metaphor for mind arrived. The sort of computer that people were thinking of is wholly governed by its hardware and programming. Surely the same must be true of us, said those who went along with the metaphor. Lots of confirmatory evidence for their view was coming from neuroscience. The more that people could use 'brain words' to describe our mental functioning, the smaller place there seemed to be for any of the old notions of free choice, responsibility and the like. By 1993, Peter Fenwick could write: 'As knowledge of brain functioning increases and imaging facilities become more available, it will become easier to detect minor degrees of brain malfunction, and the usefulness of the concept of *mens rea*, the guilty mind, will diminish even further'. His comment is especially noteworthy because he is not only a prominent British neuro-psychiatrist, but is also a leading figure in an organisation – the Scientific and Medical Network – that aims to promote spirituality in science. If Fenwick could endorse this mechanistic outlook, there must have been very strong evidence in its favour. Philosophers of course have been debating free will for many centuries, millennia indeed, without achieving much closure. Their arguments are so labyrinthine and inconclusive that we'll pass quickly over them and look at what some scientists and other non-philosophers have had to say on the topic.

People with religious affiliations were not necessarily much better off than materialists when it came to finding a place for free will. A strong current in Christian theology leads straight to Calvinistic doctrines of predestination – not much room for freedom there! Buddhists, on the other hand, tend to reject the idea that any 'self' exists capable of exercising genuine choice. The autonomous self is a sort of illusion that one must learn to transcend, they say, so how can it achieve anything as fundamental as free will? The idea of freedom in choice thus came to seem entirely dependent on experience and faith. We'll grapple with the conscious experience of choosing in the next chapter; suffice to say at this stage that 'we feel like we have it' increasingly appeared to be very shaky ground on which to base any theory that choice might actually be free. This left faith alone as 'evidence'. Quite a while before the end of the 20th century, it was looking to many as though free will and Santa Claus might be on much the same footing.

Douglas Hofstadter recalls a friend remarking apropos of this state of affairs: 'Even if we don't have free will, perhaps we have free won't'[1]. Which is actually more profound than appears. Many higher mental functions – ethics, conscientiousness and the like, presumably aspects at least of choice – depend on the frontal lobes of the brain. In fact it's been shown, since Hofstadter's friend made his quip, that these brain areas 'light up' in scans when people are making choices[2]. Much frontal lobe activity seems to stop the rest of the brain doing what it otherwise would have done, rather than stimulate it into action. So is 'free won't' a more readily available function than free will? Most people might judge it a pretty unreasonable supposition. After all, the main arguments against our having any real capacity for free will apply equally to free won't. But, as we shall see, some surprising research findings have led to serious suggestions that free won't may exist while free will does not. First, though, we need to think about what people ordinarily expect from any freedom of choice.

What is 'free will'?

This everyday feeling comes into play when we are deciding whether to go to the pub this evening or out to the cinema, or whether to take a holiday in France or go to Greece instead. Pressed, most of us would admit that these sorts of choices are often, perhaps always, either for all practical purposes random or else determined by factors of which we are not necessarily conscious. We may decide to go to the cinema instead of the pub because of a subconscious fear that John, who always makes us feel idiotic, might be in the pub. Maybe there was some especially good advertisement for holidays in Greece on the television last night that we only half remember, so we opt to go there instead of France. With more significant choices (e.g. Shall I apply for this new job? Should I marry Tom, or would Harry be better?), Sigmund Freud and the psychoanalysts have long accustomed us to the notion that realms of our minds we are not aware of have a lot of influence on what we decide. All the same, most of us would be reluctant to give up the idea that we are in some sense free to pick our course of action in a way that we notice and could recall, despite all the second thoughts and changes of mind that so commonly occur.

'Could I have made a different choice from the one I did make?' 'Yes, probably'. Unfortunately this does not really make things any clearer as far as freedom in any fundamental sense is concerned. If you have a choice between X and Y, then you may choose X. Or you may choose to choose Y instead of X. Or you may choose to choose to choose neither X nor Y. Or you may, after all, choose to choose to choose to choose X. And so on *ad infinitum*. Some unfortunate people, faced with difficult decisions, do go through mental gymnastics like this. Subjectively, the word 'choose' soon stops meaning anything when it is so often repeated. More to the point, if the straight choice between X and Y is not 'free', there is no reason to regard the second or the twentieth level of choice as any less deterministic. So the fact that you could probably have made a different decision from the one you actually made does not really throw much light on the freedom issue if, by 'freedom', we're talking about freedom from physical law – from the de la Mettrie machine.

It seems likely that what we actually want from an intuitively acceptable concept of free choice is not complete elimination of any idea that our decisions may be determined by the laws of physics and neurology, or by pure chance. After all, any such notion would be absurd. We are not 'free' to fly by flapping our arms, nor are we free to ignore the smell of baking bread when we are hungry. If bound by physical constraints in cases like these, why suppose that the situation is any different in relation to subtler 'freedoms'? Surely what we seek is assurance that our subjective feeling that we make conscious decisions, and are at least partly responsible for our actions, has *some* validity. Law and chance may govern all, but we nevertheless want to believe that we are captains of our own ship in some real sense. We may not be able to command the weather, or even have much influence over what goes on in the engine room, but we do want to feel that we can freely choose to steer north or south, east or west. This hackneyed analogy has a little-noticed implication. It is that, if you are a captain wanting to go north, you have to choose to point northwards each time you decide which way to turn the wheel. In other words, the enactment of choice, according to this picture, is extended over time and is therefore memory-dependent.

So a theory of free choice compatible with our daily lives is that it should explain how consciousness can to some extent influence its own future and the future of 'its' brain in a manner that is not wholly determined by unconscious neural or other mechanisms.

Benjamin Libet, working in California from the 1970s (most of his experimental work was done in the 70s and 80s, but he still actively contributes to debates), got some results that made it seem more difficult to develop any such view. He proved that any simple notion of conscious intention always directly and immediately causing action is untenable. This had been noted a hundred years previously by psychologist William James[3]. James famously described how, after deciding to get up from his nice warm bed on a cold dark morning, he would generally nevertheless linger in bed a while, then suddenly discover that he had risen *without* any immediately prior conscious intention to do so. Libet applied 'objective' methodology, as opposed to James' introspection, and came to much the same conclusion. People could therefore no longer dismiss it as 'merely subjective', though some did make valiant attempts, over several years, to dispute Libet's methods and thus his claims.

Libet's findings

It has been known since the late 1950s that, when people are waiting to perform some action, a small, negative voltage builds up over the front of their heads, which disappears as soon as they do whatever they are supposed to do[4]. Capitalising upon this, Libet did a whole series of studies on the timing of various types of conscious experience. In one set he told people to bend their wrists at some time of their own choosing. Meanwhile, he recorded their brain electrical activity (EEGs), and also got them to watch a fast-moving clock hand. By averaging the EEGs over the period before each wrist flex, he watched the voltage change develop. He also asked the people to tell him the position of the clock hand at the moment they first became aware of a wish to flex. He found that the electrical change started an average of 343 milliseconds – one third of a second – *before* any conscious wish to act. It therefore seemed absolutely impossible that the experienced wish to act could in any way be causing the brain activity initiating the action. In fact, it looked more like the brain potential might be causing the experience of a wish to act. Despite all attempts to discredit the finding, it has since been confirmed in several independent laboratories. Its interpretation, however, still causes a great deal of debate.

Libet himself thinks that his work conclusively shows that conscious volitions are initiated unconsciously, and thus by implication cannot be 'free'. Most commentators seem to agree with him about the unconscious initiation of voluntary action, though not necessarily about the 'freedom' implication. However, Libet also points out that the actual wrist movement occurs on average 550 ms after the onset of the potential in the brain. In other words, wrists are not bent until after the wish to act has become conscious. The time it takes for nerve impulses to travel from brain to arm muscles – 50 ms – can be ignored. Nevertheless there is still a window of around 150 ms (i.e. 500 ms minus 343 ms) in which consciousness could veto any incipient, unconsciously initiated action. Libet argues that free will may in fact be free won't, operating within this window of opportunity. He has suggested that consciousness is a 'mental field' that is free in some sense to arrest deterministically initiated actions.

Few have shown enthusiasm for the 'mental field' idea. Free won't but no free will is also generally regarded as dubious. After all, if there is unconscious initiation of 'voluntary' action, there is every reason to suppose that similar unconscious initiation of voluntary inaction exists, though we have not yet developed the right experimental setup to detect it. Debate has tended to centre on the idea that, although the unconscious minds of Libet's experimental subjects chose exactly when to flex their wrists, the people had already 'consciously' chosen to participate in the experiment and to act at *some* time during experimental sessions. Although the tactical timing of the 'voluntary' choices was not initiated consciously, it has often been suggested, the general strategy of participation may have been subject to free choice. Objectors to this view say that strategic choices, too, are probably initiated unconsciously, so are unlikely to be any more free than tactical ones. Supporters point out that there are obviously, in the real world, very complicated reciprocal relationships between conscious volitions and their unconscious roots, which are hard to pin down in artificial experimental settings. There may well be a place, they say, for some sort of genuine freedom of conscious choice within these complex feedback loops. It's been a vigorous debate, but one that has often generated more heat than light.

The physics of free will

As de la Mettrie would probably have said, there is not much room for free will in Newton's clockwork universe, where the future of everything is exactly determined by its initial state. The advent of quantum and, to a lesser extent, chaos theory changed all that. Quantum theory altered our view that everything follows inexorably from how it begins and deterministic chaos put the tin hat on predictability. Many people have therefore sought a home for free will somewhere within the bounds of these two theories. The main difficulty is that both theories substitute pure chance for some of the determinism of a Newtonian universe. And chance is actually no more free will-friendly than determinism. However, it is worth taking a quick look at what quantum theory in particular has to offer, if only because it is our most fundamental theory of how matter behaves, and this includes the matter in our brains.

Henry Stapp, a physicist based at the prestigious Lawrence Livermore laboratory in California, has made what is probably the most thorough attempt so far to build a theory of free will from fundamental physics. He's been working on it since the mid-1990s. He points out that, when we observe a quantum system, two 'choices' determine what we shall see. The first is the 'Dirac choice'. This is strictly random and refers to the probabilities described by the quantum wavefunction. The wavefunction shows that, if you look for where a particle is situated for example, there is a certain probability that you will find it here and another probability that you will find it over there. The Dirac choice simply refers to the 'choice', attributable in a sense to the particle's wavefunction, between whether the particle itself will actually be here or there when you see it. There is no room for free will in the Dirac choice because it is entirely random for all practical purposes. There are actually some much disputed claims that consciousness may be able to influence random events[5], but the effect is extremely weak, even if genuine. Moreover, if it happens, it cannot be accounted for in terms of any present-day physical theory, not even current quantum theory. The deviation from strictly random behaviour that has been claimed involves only around one in five to ten thousand events. Even if one accepts that the effect really does exist, it is apparently so feeble that one cannot see how it could possibly provide an adequate basis for everyday free will.

The second type of choice, the 'Heisenberg choice', looks a bit more promising as a possible 'home' for free will. It refers to which aspect of the wavefunction you choose to observe. For instance, you may choose to see where a particle is situated or to measure its speed. Heisenberg's Uncertainty Principle tells us that the more information you have about a particle's exact position, the less you can know about its speed and vice versa. Similarly, if you choose to measure the spin of a particle in the vertical direction, you lose all information about its horizontal spin. 'Observers' make these Heisenberg choices of what to observe, but what are the observers themselves? Well, they are conscious people in the traditional Copenhagen interpretation of quantum mechanics[6], but nowadays they are usually regarded as any system, conscious or not, that interacts with another resulting in 'decoherence'[7]. The Heisenberg choice does not refer to any sort of conscious choice, therefore, but only to the aspects of two systems that happen to interact. A particle that bounces off another, for instance, will have 'chosen' to measure the momentum involved. An atom that absorbs a photon will have 'chosen' to measure the position of the photon.

So, despite his own liking for the Copenhagen outlook on quantum theory, Stapp could not use the notion of Heisenberg choice in any direct way to provide a home for free will. Nevertheless, he thought it could be at the basis of free will if there exists some mechanism that meets two requirements: first, a means by which the conscious brain could 'voluntarily' bias the Heisenberg choices that it was necessarily making all the time in huge numbers; and second, that the mechanism should be able to affect its own future evolution considered as a quantum mechanical system described by a wavefunction. Although we usually think of wavefunctions as associated only with single particles, everything has one, including brains and all their sub-systems. These functions evolve in an entirely deterministic way except when decoherence happens and random Dirac choices occur.

Stapp's first requirement was simply met. Attention biases our Heisenberg choices. If we look in one direction, for example, we 'choose' to measure the positions and/or energies of photons coming from that particular direction rather than another. It was the second requirement that presented the difficulty, for if the attention system could not affect the future evolution of its own wavefunction then there would be no place in it for freedom of choice; all would remain as deterministic or as random as ever. Stapp appealed to the quantum Zeno effect, which can be caricatured as 'a watched pot never boils'. If a quantum system is 'observed' very frequently, its wavefunction won't have time to evolve between observations, so it will, in effect, get stuck in one state. Stapp suggested that attention might induce a Zeno effect, thus influencing its own future evolution and allowing it to select its Heisenberg choices according to its own 'wishes', a possibility which would indeed provide a sound basis for free will. There is a problem, though: a discernable Zeno effect requires the 'observer' to be checking millions of times per second. It is very hard to imagine how any physiological system as extensive as that underlying attention could pull off anywhere near the necessary observation frequency.

Many people have tried to introduce 'freedom' into the workings of the brain at the level of fundamental physics, often in online discussion groups such as *Quantum Mind*, which is sadly now defunct[8]. All such attempts have either involved appeals to some sort of new physics that we don't know anything about, or invoked neurophysiological implausibilities. Freedom is probably best sought at less fundamental levels until we have a clearer idea of what sort of thing it might be.

A psychologist's tale

On the whole psychologists, perhaps wisely, have left issues of 'freedom' to one side and have concentrated on mind-brain interactions. One of their underlying assumptions seems to be that, if the conscious mind can affect the workings of the brain, then something closely akin to most people's idea of free will must exist. Max Velmans is a London-based psychologist who has led many of the recent discussions on this topic[9].

He has drawn attention to a number of issues that have to be taken into account in any sensible view of how mind and brain relate to one another. The first of these is that mental states, including conscious mental states, can certainly affect the *body*. The whole of psychosomatic medicine is based on such effects, from relaxation therapy, through changes that can be induced by hypnosis, to placebo effects. Some of these effects can be very dramatic. For instance, a proportion of hypnotic subjects produce skin blisters purely through suggestion. Placebos can reduce swelling or wheezing, as well as subjective feelings of pain or illness. If mind can so affect the body, it would seem absurd to suppose that it does not equally affect the brain, since the mind is more intimately involved with the brain than with the rest of the body.

A second issue to which Velmans drew attention is what might be called the Oscar Wilde phenomenon, after the line in one of his plays: 'How should I know what my opinions are until I hear myself speak?' When we do anything – pay attention, go for a walk, write a letter or whatever – we formulate some general conscious strategy like 'I am going to sit down and send an email to Auntie Mabel', but we have no conscious knowledge whatsoever of the details. We never know how our motor cortex moves us to the right place, how our attentional apparatus focuses on the keyboard, or where the words come from that we type. All that is entirely unconscious and automatic.

To return to Libet for a moment, he showed (in a whole range of experiments on the timing of conscious experience) that we always live at least one fifth of a second behind the times, because it takes 200 ms or more for us to become conscious of information arriving at the brain. The brain somehow compensates for the time lag, so we seem to ourselves to be living in the present. This compensation provides another example of how some of the content of consciousness is illusory, like the filling in of the blind spot that so impressed Daniel Dennett (see Chapter 2). It is worth noting that the delay in consciously perceiving information is entirely consistent with my claim that consciousness is intimately entangled with memory. If information were to reach consciousness as soon as it got to the brain, there would be no time for memory to develop and consciousness could not be regarded as memory-related. As it is, Libet's findings, surprising though they seemed to many at first, are exactly what we would expect. The actual figure for the time it takes for the relevant memory to develop (over 200 ms) lends support to the view that consciousness may be most closely related to the stage of the memory process at which information is being edited for inclusion in working memory.

Clearly a lot of what goes on in relation to both the psychology (Velmans) and the neurophysiology (Libet) of apparently 'conscious' perception and behaviour is in fact unconscious. Consciousness is, it seems, the tip of an iceberg. Responsibility for the detail of what we do lies with the unconscious bulk of the berg.

In 2002 Velmans summarised his conclusion thus:

> Viewed from a third-person perspective our own preconscious mental processes look like neurochemical and associated physical activities in our brains. Viewed introspectively, from a first-person perspective, our preconscious mind seems like a personal but 'empty space' from which thoughts, images and feelings spontaneously arise. *We* are as much one thing as the other – and this requires a shift in our sensed 'centre of gravity' to one where our consciously

> experienced self becomes just the visible 'tip' of our own embedding, preconscious mind.

There is much that is both wise and true in this. It provides a believable account of the origins of choice, but leaves little room for *free* choice. Consciousness emerges from Velmans' preconscious 'brain machine' – and machines, of course, are far from free. The notion of voluntary choice may indeed have to be extended to include preconscious 'choices', but this would seem to put us firmly back in de la Mettrie territory.

The judge's verdict

One of the most attractive features of the subject of consciousness studies is that it cannot be confined to philosophers, neuroscientists or other specialists, since it encompasses the whole of human experience. David Hodgson is an Australian Supreme Court judge who, perhaps for obvious professional reasons, wants to find adequate grounding of some sort for notions of freedom and responsibility. He has come up with a number of ingenious arguments. One of them, written in 1999, points out that the belief that the evolution of absolutely every system is wholly due to causal laws or to chance goes back to the Scottish Enlightenment philosopher David Hume, writing in 1748 (the year in which his *Essay Concerning Human Understanding* was published). Hodgson suggests that it is in fact incorrect. He pictures three different types of universe. The first is a 'chess universe' in which there are rules, but also a guiding hand that selects what shall occur. Few people nowadays go along with this Cartesian picture. Next, Hodgson sketches a 'life universe' in which everything that happens is due to rules analogous to those governing cellular automata, maybe with the addition of a random component. These automata have been wonderfully well described by Stephen Wolfram. They are computer programs which produce ever-changing patterns according to rules which may be very simple, but can give extremely complex results. In fact, Wolfram has shown that automata themselves can implement any possible computation (although it might take them an extremely long time to actually complete some particular computation). This 'automata world' is the sort that de la Mettrie envisaged, in which many of us nowadays still think that we are living. It is also the one in which Hume's assumption holds true.

A third type of universe is conceivable. Hodgson calls this the 'superlife' one. Here there are deterministic and statistical rules, just as in the automata, 'life' world, but there are also integrated assemblages of particles. Each of these is unique and, crucially, retains *memories* of its own previous states. The evolution of this universe conforms to rules, but there are nevertheless a variety of pathways that each assemblage can follow. Which particular trajectory it will take is determined by its choice of pathways at various times. In a superlife universe, each assemblage could certainly be regarded as responsible for its long-term behaviour. If its choices between alternative pathways were partly dependent on its memories as well as on its immediate circumstances, it would also possess a faculty that is equivalent, for all practical purposes, to our working concept of free will. If the captain of the ship consciously remembers which way he wants to go, and turns the wheel accordingly, then he has what amounts to a free pick of direction. Since it is fairly obvious that the 'superlife universe' is the sort of world that we *do* inhabit, Hodgson concludes that both responsibility and free will are meaningful concepts. Those who think that the world we actually inhabit is Hodgson's 'life' universe have forgotten to account for memory.

Next Hodgson goes on to look at the specific part that consciousness might play in choice. Consciousness embodies information about the qualities of things, and also gathers information that is represented in separate parts of the brain into unified experiences. It therefore enables people to make global judgments, he argues – especially aesthetic judgments – about what is going on. These judgments influence the choices that we make in a 'superlife universe', so allow consciousness to be responsible to some extent for decisions. If consciousness relies on memory, it can also be regarded as possessing freedom of choice in the sense that its own past will have affected the nature of its present choices independently of any non-conscious influences. Since we have already seen that consciousness is indeed intimately entangled with memory, it is clear from Hodgson's argument that we do possess freedom of this type. It seems a pity that his views have attracted less attention than might have been expected from the force of his arguments.

The steam whistle

T. H. Huxley, the great Victorian biologist known as 'Darwin's bulldog', believed that consciousness is epiphenomenal. He likened it to the whistle of a steam locomotive that contributes nothing to the running of the engine. The proposal has recently and rather surprisingly been revived by a neurophysiologist from New Zealand, Sue Pockett. Her views provide a particularly clear and explicit example of an undercurrent that pervades some scientific thinking about the place of consciousness.

Pockett contends that consciousness *is* a particular, though currently unknown, type of pattern in the brain's electromagnetic field. Any neurophysiologist might be expected to view such a field as likely to have effects of some sort, the precise effects being dependent on the nature of the pattern and its strength. Pockett, however, thinks of it as a bit like the ripple reflections on the ceiling of an indoor swimming pool. There's no way the pattern on the ceiling can affect what goes on in the pool. It is not at all clear, though, why she should attribute 'consciousness' to the pattern but not to the neural processes generating it (which certainly cannot be epiphenomenal). Presumably she thinks of her 'consciousness pattern' as some sort of emergent property which is distinct and different from the processes underlying it.

If consciousness is distinct from what gives rise to it and has no effect upon its trigger, there seems to be no way it could get remembered. But everyone can remember what it is like to see a sunset, for example. In defence of her position, Pockett suggests that, when we recall a sunset in our mind's eye we are re-creating it from unconscious memories, which is true enough. The next step in the defence is not so true, though. It is the argument that, since we re-create this sort of memory each time, we don't remember our *consciousness* of the sunset as such, we only remember some of the neural activity that originally generated the experience. This claim does not work because it is perfectly possible to remember the fact that we saw a red sunset, as opposed to an orange one for example, without having actually to picture the red in our mind's eye.

A diehard Pockett fan might go on to claim that what we are re-creating is something that we said to ourselves at the first viewing of the sunset. Most of us would probably agree, though, that it is possible to remember that a sunset was red *without* having said to ourselves at the time of viewing: 'This is a red sunset, not an orange one'. More tellingly, maybe, no one has any problem remembering it if they felt a twinge of toothache yesterday. They don't have to re-create the pain in their imagination (it is actually very difficult to picture a pain to oneself in the same way that one can recall a shade of red or a tune). Nor is it necessary for them to say to themselves at any time 'I've got toothache' in order to be able to remember. Even when based on a concrete model like Pockett's, it is clear that the claim of epiphenomenalism amounts to no more than a semantic quibble, in this case involving an artificial distinction between an electromagnetic field and what generates it instead of the distinction between consciousness and its neural basis proposed by some philosophers (see Chapter 2).

Genetic determinism

You'd have thought that genetic questions would be central to any discussion of free will, given the furore that arose when leading US biologist E. O. Wilson published his books *Sociobiology* in 1975 and *On Human Nature* in 1978. In these he argued that many of our characteristics and actions are due to our genes. Certainly there is always a lot of media excitement whenever anyone claims to have discovered a gene 'for' criminality or homosexuality or whatever, and it is often implied that the people with the alleged gene haven't much option about how they will behave. But any genetic constraints on choice are no different in principle from other physical constraints. Someone born colour blind is never going to be free to become a really great oil painter, whatever his talent for drawing. Mothers generally have little choice about whether they will love their babies, since all sorts of hormonal mechanisms, ultimately based on their genetic make-up, ensure that they will usually bond successfully. Similarly, some people's genetic make-up may predispose them to anti-social or other particular types of behaviour, but their choices within this constraint may be just as 'free' as those of the rest of us in relation to our mostly law-abiding lives. Genetics sets limits on what we *can* do, and maybe often biases the probabilities when it comes to what we *will* do[10] – but that's all.

Y

Things are still looking a bit grim for any user-friendly concept of free will, but not nearly so bad as they did at the start of this chapter. All the views we've just looked at either explicitly or implicitly involve memory in theories of choice, though in very different ways. If memory is there, so too is consciousness. And if consciousness is involved in choice, freedom of some kind cannot be far away.

Hodgson's proposals explicitly give memory a pivotal role in the introduction of responsibility and freedom. Stapp, on the other hand, proposes that sustained attention is the crucial factor. But sustaining attention is impossible without memory, so this is implicitly central to his theory too.

In Libet's case, too, the involvement of memory is implicit. The delay between the neural events preparing for action and the conscious awareness of wishing to act hint that the latter is a very short-term memory of the former. This is consistent with his other findings about the time needed for conscious registration of information arriving at the brain[11]. The delay has usually been taken as an indication that consciousness does not play a role in causing action, at least in tactical voluntary choice situations like deciding when to move one's wrist. However, because consciousness is memory-related, delays are inevitable if it is involved in these choices. Velmans has provided an insightful account of how its involvement might work. Finally, the 'epiphenomenal field' picture of consciousness is a neat, albeit inadvertent, demonstration of how ignoring memory altogether simply doesn't work. So where does the other axiom fit in – i.e. the view that all unique conscious experiences are accompanied by unique neural events of some sort?

Chapter 5

ILLUSIONS OF WILLING

Whenever most of us reach out to pick up a paper, or think about what we would like to do tomorrow, we feel that it's *us* who have done these things. Equally, if we see someone else eating a hamburger, we don't feel that we are responsible for his or her actions; not even if our mouth waters and we feel an urge to do the same. Not so for everyone, though. People with schizophrenia commonly feel that their thoughts have been put into their heads by some outside agency or that their actions are not their own[1]. They can also experience 'delusions of influence': they feel themselves responsible for the actions of strangers across the street, or the behaviour of passing traffic, or even for the weather.

There are a great many theories about the causes of schizophrenia but, whatever they may be, the results are inappropriate activation or inactivation of particular mental faculties. These faculties may be rather general. For example, people may lose the ability to feel much emotion of any sort, or they may experience a tremendous increase in their suspicion of everyone and everything. On the other hand, the phenomenon may be relatively specific. Sufferers of 'Capgras syndrome' say that their relatives and friends, although looking just the same as always, have actually been replaced by strangers[2]. The disturbance is usually attributed to some sort of disconnection of the neural centres responsible for recognising faces from those holding memories of affectional ties. People still recognise the faces, but they don't feel the emotions and other memories that should go with them. Without the sense of familiarity, they conclude that the faces must belong to strangers.

It is becoming ever more apparent that, wherever there is a mental faculty, researchers will probably find a neural module that underpins it. These modules can be temporary functional assemblies of neurons or can be more permanent, anatomical structures. Judging by how narrowly localised within the brain many mental functions appear to be, they are probably often permanent or semi-permanent. Some researchers at present believe that 'module' refers only to genetically pre-programmed, hard-wired faculties, but this is unduly restrictive. Learning is needed too, and some modules result solely from learning[3]. In relation to conscious mental faculties, the existence of such specialised modules follows directly from our first axiom, outlined in Chapter 2: they are needed to embody the distinctive neural properties of the distinctive components of conscious experience.

Normally separate modules work together to give us an apparently seamless flow of experience. In schizophrenia, though, there is quite literally a lack of joined-up thinking. Gaps appear, and the patchwork nature of mentality becomes visible. Alternatively, individual modules can go into overdrive in this terrible illness, and produce experiences that most of us are spared: for example, a feeling, on glancing through the window, that the particular number plate on a passing car means that something terrible is going to happen very soon, or a conviction that, if the traffic lights turn red before the car gets through, then aliens must be in control of the city. Actually the 'overdrive' phenomena may well be due to the same factors as the 'seams falling apart' ones. Just as activity in some separate modules may fail to join up properly in experience, so modules may become severed from centres that would normally prevent them from over-working inappropriately.

There are specific neurons which fire when a monkey sees a particular face or when a rat reaches some particular place in a maze that it recognises. Such cells used to be called 'Grandmother neurons' (i.e the specific neuron that lights up when you clap eyes on your grandmother). The name was a sort of joke because there was scepticism about their existence, but any scepticism has now gone. (At the time of the Monica Lewinsky affair, it was claimed that researchers had found a 'Bill Clinton neuron' in some people – that may be an urban myth, but the other findings are indisputable.)

In a similar vein, researchers frequently rediscover that people with epilepsy affecting their temporal lobes may have religious experiences, and that such experiences can be induced artificially by stimulating particular parts of the temporal lobe[4]. There has also been much interest in the quite recent discovery of 'mirror neurons' in the pre-motor areas of our own brains, as well as monkey brains, which become active simply when we watch someone else doing something such as lift a ball or smile. The same cells seem also to be involved when it comes to actually performing the actions ourselves. They're not just confined to vision, either. Some of the neurons involved in the act of tearing paper also light up at the sound of paper tearing. It has been hypothesised that mirror neuron systems underlie human faculties for mimicry and perhaps even for empathy.

Responsibility detection

Given this background, it would not be too surprising if the *feeling* that we are consciously responsible for something were also a function of some specialised module. Experiences do usually give us valid information about the world or ourselves. If you're walking along and see a car coming towards you, there's an outside chance that it is an illusion or a hallucination. All the same, there probably really is a car out there and it is wise not to cross the road just now. If you feel your stomach churning after a meal, you may well have eaten something that disagrees with you and it is best not to stray too far from the nearest bathroom. Similarly, when we have the feeling that we are responsible for an action, it may well usually be true that we *are* responsible. It is not necessarily true, however, that the module responsible for the conscious feeling is the same as the module responsible for the action. The two functions might not be closely connected at all. In fact there's quite a lot of evidence that they are indeed separate.

On the whole it seems to be easier for people to feel they are not responsible when they are, rather than the other way round. The Harvard psychologist Daniel Wegner described all this magnificently in his 2002 book *The Illusion of Conscious Will*. Every day, all over the world, people involved in cult religions get into states where they feel 'possessed by spirits' and no longer the authors of their own behaviour. Wegner quotes a lovely anecdote (which may be apocryphal) about the ventriloquist Edgar Bergen, whose dummy was called Charlie. A visitor came into the room one day and found Bergen having a conversation with Charlie:

> When Bergen noticed that he had a visitor, he turned red and said he was talking with Charlie, the wisest person he knew. The visitor pointed out that it was Bergen's own mind and voice coming through the wooden dummy. Bergen replied, 'Well, I guess it ultimately is, but I ask Charlie these questions and he answers, and I haven't the faintest idea of what he's going to say...'

In experimental situations, too, people can be fooled into attributing responsibility for what they have done to others. Subjects asked to press keys as a way of answering questions, and given a 'helper', can be induced to attribute correct responses to the helper (who in fact does nothing). More relevant to day-to-day life, it's quite common for those involved in some group activity to attribute responsibility for their own actions to the group, not themselves. Findings like these could be taken to indicate either that action and awareness of responsibility for it are quite separate functions, or that they are basically the same function but the awareness component can sometimes get switched off.

Feeling you are responsible when you're not does happen in normal people as well as psychotics. Indeed it seems to be something of a personality trait in some people. Many a long-suffering mother blames herself for everything that has gone wrong in her family; lots of conscientious workers feel far more responsible than they should for the actions of others. In experiments, people have been induced to remember that they were guilty of doing something bad, when in fact they were entirely innocent. But these examples raise rather different issues to do with character traits and long-term memory, and all sorts of additional complications.

Straightforward misattribution of responsibility to oneself can be produced experimentally in situations which have nothing to do with being an over-conscientious worker or spouse, or developing a false (declarative) memory. One experiment, for example, involved two people apparently collaborating in moving a pointer on a computer screen by jointly controlling a mouse. Unknown to the actual experimental subject, one of these people was really an experimenter who was sometimes getting instructions to make particular mouse movements. However, the subject felt responsible for these movements (which in fact had nothing to do with her) if there were certain temporal or informational constraints on the relationship between her own intentions and the actual pointer movements. The brain assesses its own responsibility from external as well as internal cues. It is *not* the case that acting directly results in a sense of ownership of the action. There are two separate functions, which usually, but not always, work seamlessly.

So, to recap, feeling you're not responsible when you are shows that action and feeling responsible for it are either separate or they are basically the same, but the feeling component can get switched off. Feeling you're responsible when you're not whittles it down, because it shows that action and and feeling responsible must be separate.

Why should we need an 'ownership of responsibility' module? Well, the main function of the brain from a Darwinian point of view is to predict what could happen from present circumstances, and then try to arrange things so that the foreseeable future includes the essentials for survival and mating, such as eating rather than being eaten. Prediction is possible only if the brain has a very good idea of what it is, and what it is not, responsible for. If a foot moves, for example, it could be vital for the brain to know whether it had moved the foot, or if the movement was due to something slithery that the foot is standing on. A 'responsibility for action detector' is an essential part of the toolkit of any brain capable of survival.

What can hypnosis tell us?

The question of whether the 'responsibility detector' is right or wrong in its estimates in people who have been hypnotised[5] raises all sorts of conundrums. It opens a door to issues of central importance later on – problems to do with how much of our freedom is in ourselves and how much in our circumstances and societies. A stage hypnotist can tell a good hypnotic subject that he or she has changed into a domestic robot and will kneel to polish shoes with their hair at the click of his fingers. Or at least it used to be possible in the less PC, Barnum and Bailey, days of showmanship. Be that as it may, suppose the instructions are given, the fingers are duly clicked and the subject acts in the way suggested. When asked later why they did so, they usually say something like: 'It wasn't me', or 'I felt compelled', or 'I don't remember doing that'. Yet it was clearly the subjects' own brains that carried out the action. Their 'responsibility detector', in the first two cases at least, appears either to have switched itself off, or to have decided that the 'real' responsibility lay with the hypnotist. There's also a possibility, of course, that the subject's statements are simply a form of confabulation, like that produced by Korsakov patients, arising because any output from the 'responsibility detector' didn't get into declarative memory in the normal way. Nevertheless, the question still arises, supposing the detector did judge that the hypnotist was responsible, was it right to do so? If it was correct in some sense, what does it mean for a brain not to be responsible for actions that it carries out?

The opposite shift in attribution can also occur. Our stage hypnotist can easily leave his subject with a post-hypnotic suggestion that, after waking up, the first time the hypnotist says the word 'abracadabra', the subject will jump on the table and sing *God Save the Queen*. The hypnotist clicks his fingers to 'wake' the subject, then turns to the audience and starts talking about magic. Lo and behold, as soon as he says 'abracadabra', the person jumps on the table and sings. This time, he or she generally[6] *will* claim responsibility for action. If asked why they should do anything so bizarre, they will usually produce some rationalisation such as: 'I thought we had got to the end of the show and it was the right time to sing the National Anthem'. In this case the 'responsibility detector' does appear to have been switched on, and to have reached the opposite conclusion. Korsakov-type confabulation cannot be blamed here because the subject is fully 'awake' at the time of singing.

Interestingly, though, hypnotists wishing to induce post-hypnotic actions often tell their subjects that they will remember nothing of what went on while they were 'under'. Maybe, when subjects claim responsibility in these circumstances, they do so because information about the post-hypnotic suggestion is not available to their 'responsibility detectors'; it has either been genuinely forgotten, or at least repressed and made unavailable to the relevant brain areas. Unfortunately, there does not seem to be any quantified information about whether people still claim responsibility for actions of this sort when they *haven't* been instructed to forget their instructions. There is anecdotal evidence, though. Alan Gauld, an expert on hypnotism who worked in Nottingham University's psychology department, comments that in these circumstances, 'If the subject recollects the suggestion when the time for fulfillment arrives, he is likely not to carry it out: if he does carry it out, he is likely to say simply, "You wished it so I did it"'. In other words, if in these circumstances the detector has information about a suggestion that the person has acted on, it may claim immediate responsibility, even though 'higher level' cognitive processes of some sort attribute ultimate responsibility to the hypnotist. Whatever the implications for the involvement of memory in these judgments, at least these considerations help to confirm that the 'responsibility detector' is a separate module that makes its assessments from information gathered from a range of sources. The centres that actually perform the actions are only one among several such sources and therefore cannot themselves be directly responsible for the conscious feeling of ownership of action.

We agreed, on the basis of the first axiom, that the conscious feeling of ownership of an action must be associated with special neural activity of *some* sort. All distinctive conscious experiences are associated with distinctive neural activity. There was no reason to suppose that the experience of feeling responsible for an action would be an exception. What we did not know was whether the neural activity in question was some aspect of the *same* activity that actually generated the action, or whether it was separate in some way. It might have been either because only a subset of all the neural activity going on at any particular time is associated with conscious experience. There was no *a priori* reason to suppose that action-producing neural activity should be part of that subset.

We have now seen that the conscious feeling of owning an action does seem to be associated with activity in a separate 'responsibility detecting module'. The feeling represents the brain's best guess about when it has itself caused something, and the guess can sometimes be wrong, in either direction. Responsibility can be felt when there isn't any, and not felt when there is. In fact, there is direct experimental evidence that the module really is separate. It's not all a matter of inference. The feeling of being in control of one's actions, which is closely linked with (though not necessarily exactly the same as) the feeling of agency, is an attribute of a module that predicts the expected *future* outcome of actions, subsequently checking from sensory feedback that what actually happens matches expectation. This module is known from scans to be sited in brain areas distinct from those that directly cause movement. There is every reason to suppose that the ownership of action feeling 'I did it', being so similar to 'I'm in control of it', is either wholly or partially a product of the same module.

Some people, Sue Pockett with her epiphenomenal field picture for instance, take considerations of this sort as support for the idea that consciousness itself does nothing. Others argue that it must indicate a lack of freedom in choice. If the feeling of being responsible for choosing is no more than yet another function that the brain computes, they say, no real 'freedom' can be associated with it. Of course, simple evolutionary considerations suggest that the brain's feeling that it has 'chosen' an action must usually be right. We would not be here to tell the tale if it wasn't, because knowing whether events are due to one's own actions or to an outside agency is essential to survival. It is the neural computation of ownership of action that matters, not the conscious feeling, say the group who argue that consciousness itself does nothing. The others concede that consciousness may be identifiable with aspects of a computation. But, they point out, that the computation is usually correct does not mean that it is 'free'– rather the opposite. Our awareness of 'free choice' is indeed something of an epiphenomenon, both groups agree, reflecting the outcome of a quite deterministic computation that provides a sort of read-out of the unconscious neural mechanisms actually responsible for actions. That would surely be the most reasonable conclusion to draw from the first axiom alone, in the light of the psychological evidence.

Luckily, though, there is also the second axiom; the one that entangles consciousness with memory, and in particular with that part of the memory process that edits information for inclusion in working memory. If free will is taken to mean the idea that consciousness can influence its own future and that of the brain to which it belongs, independently of any unconscious neural or other physical causes, then the role of memory is crucial. As it turns out, the *actuality* of free choice could never directly produce the *feeling* of free choice. The feeling relates only to timescales of a few seconds at most; the actuality relates to the timescales embodied in memory, which range from under a second to decades or even a lifetime.

The tale told in the next chapter illustrates the part that memory plays in giving consciousness a degree of autonomy in relation to the unconscious processes that underpin it. It is not the full story about free will, because, as we shall soon see, there are also all sorts of social and cultural constraints on this, which have no direct relationship to physical or neural constraints.

Chapter 6
SUSAN'S TALE (PART I)

The year is 1958 and Susan is just two days old. She is lying in her cot, drifting in and out of sleep. Awake, she can see all sorts of blobby shapes, wandering in and out of her field of vision. Some special ones attract her attention. In fact she can feel herself drawn to them. She does not know who 'herself' is, but she is nevertheless conscious of acting in relation to these special objects, which she will later learn are faces[1]. There is no real choice here, though, since babies are hard-wired to attend to faces. Somehow their brains develop so that, right from the moment they're born, they will spend more time looking at faces than at other objects in their field of vision. So it is unlikely that Susan would have been able to ignore one at that age, despite any feeling that she might have of choosing to look. A particular face is more interesting than all the others. It has already been stored in Susan's memory, even though she is so young. She will always be able instantaneously to recognise Mummy's face, which goes along with Mummy's voice and smell, and with all those yummy feelings you get when your tummy is filled. With Mummy's face in her memory, it is possible to spot the differences when she sees others. She can begin to learn to recognise them as well.

A few weeks later she is able deliberately to reach out for things and sometimes to grab them. Something a bit more like real choice is becoming possible. If there's a furry teddy bear and a shiny plastic ball in front of her, she may decide to go for the plastic ball instead of the teddy, even though she can't name them to herself yet. There's no conscious deliberation about which she would prefer. It is more that she just feels an urge to go for one or the other and acts on it. She has already learned that, if you try to get both at the same time, you don't succeed.

It is not too long before she is quite mobile. She can crawl, but mostly gets about using a bottom shuffle technique that she has developed. She sits looking like a small Buddha, then twists a bit and does something with her legs – and there she is somewhere else. Mobility has greatly expanded the range of choices open to her. You can sometimes see that she is consciously weighing in her mind questions like, 'Should I go this way for that thing with a face on it (a doll), or would it be better to go that way and play with the one that rolls along (a cart)?'. Maybe, when she chooses the cart, an adult might say, 'That's because she played with the doll most of the morning, so she is bound to want a change'. Susan doesn't see it like that, though. She has decided to go for the cart and that's all there is to it. But there are other things that she can't decide about doing or not doing. One of these, something that she has to do whether she likes it or not, will soon result in an even greater enlargement of the possibilities open to her.

Just as babies are hard-wired to attend to faces, so too they probably have to attend to voices, and especially to Mummy's voice when she is doing baby talk. Though it is not known how they do it, they have an in-built capacity to make sense of grammar, regardless of what language is in use around them. So, willy-nilly, Susan learns to understand what is said to her and to speak[2]. Now there are all sorts of stories that she can choose to hear: The Three Little Piggies, Goldilocks, Sleeping Beauty, Cinderella, and many, many more. She can decide which she likes the best, and which ones she wants to hear again. As Mummy soon discovers, she has an apparently inexhaustible appetite for repetition of favourites. In consequence, lots of new ideas and concepts open up before her and get firmly lodged in her mind. Also, saying things brings with it endless possibilities for choosing what you will say and for trying to influence what is going on around you. Of course, if you want to influence something, you have first of all to decide what it is that you want to affect. In Susan's case the target is often her older sister Diana, either to try to get her to play or to try to make her stop being annoying.

She is actually quite a lucky child. Daddy has a good job as a technician, and Mummy has elected to stay at home to look after the two children, at least until Susan goes to school. Both parents have the liberality and sense of endless possibility that was so prevalent in the early 1960s. But they are sensible folk, even a bit boring some would say. They are not into drugs or spouse swapping, or indeed anything likely to upset their children. Susan has good, safe boundaries within which she can exercise a wide range of choice. So it is that her own conscious decisions have a lot of effect in moulding her character. For of course each time she attends to some particular thing, she is excluding a wide range of alternatives. The older she gets, the more her own consciousness determines what will get stored in her memory and what won't.

The cumulative effect of her choices is considerable. For instance, she is a child who likes detail. She will cuddle up to Mummy when she is doing embroidery and will watch with fascination how each tiny stitch adds to the growing pattern. She will spend minutes on end (a lifetime to a child!) seeing how ants in the garden go about their endless tasks. If Diana calls her, she deliberately does not answer at first since she has discovered that it is never a good idea to give in to Diana too easily. She is becoming someone who, among much else, will always be interested in minutiae, and one who knows her own mind. Given the choice, she will often go through one of her picture books rather than play with Diana's doll's house. Perhaps as a consequence of such decisions, she will never want to become an interior designer or the like. Her own consciousness has always had a large influence on her immediate future, due to its links via short-term memory with attention. It is now already beginning to affect the sorts of choice that she is likely to make in the quite distant future as well.

In next to no time, from her parents' point of view if not from her own, she is in her first year at school. One day in the playground, she hears Hannah making fun of Rosie. Rosie is wearing a hair band that looks as if it is made from an old dish cloth, and Hannah is keeping on about it: 'What's that thing? I think it's a dead rat! Your Mummy must be stupid! Why can't she get you a proper band?...'.

Susan notices Rosie's eyes beginning to fill with tears, and remembers how she herself feels when Diana won't stop getting at her. She goes over to Rosie and gives her a hug. It is more of an impulse to act than a very deliberate choice, though Hannah is a bit scary so she does feel a moment of hesitation. 'Go away', she says to Hannah. 'You're not nice'. Surprisingly, Hannah does go away. She pretends a sudden interest in a game going on across the yard and goes off to join it. A teacher sees what went on and later, in the corridor, says to Susan, 'That was a kind thing to do for Rosie. You are a good person, Susan'.

'So I'm a good person'. says Susan to herself. 'I wonder what good people are like...'.

Saturday afternoons in the winter can be a bit scary too, though not in the same way that Hannah is scary. After tea at the kitchen table, everyone settles down in the living room to watch *Doctor Who* on the television. There's this old guy who travels in a telephone box that is really a time machine, saving people from nasty, nasty things called Daleks, which are all made of metal or plastic or something. They like to *kill* people. They point this stick thing they have at people, and the people turn transparent and then fall down *dead*. When the Daleks appear, Susan cuddles up to Daddy and sometimes can't even bear to watch. The Doctor usually has a lady with him, too, who often needs rescuing but sometimes does sensible and brave things without being told. She's a good person, and so is the Doctor in spite of his funny ways at times. Susan is forming a whole range of ideas about what it is to be such a person.

What with lessons and friends and sports and books and holidays and hobbies, she sometimes feels her head is so full that it will burst if anything else goes in. Nevertheless, as she discovers in 1972, there is room for more. Dave has arranged to meet her behind the school bicycle shed – that perilous trysting place of which many stories, likely and unlikely, have been told. For some time she has been feeling giggly and kind of melting whenever she sees Dave. Were the word current then, she would have described him as cool. She does not connect these feelings with the sex education lessons provided at school. Anyhow, there she is behind the shed and Dave is kissing her. He's a bit clumsy, it has to be said, but nevertheless it feels dreamy and nice. Then his hand strays up her leg. Looking over his shoulder, she notices a particularly shiny bicycle saddle. It reminds her of Hannah. Susan has several times overheard boys snickering about the school bicycle – the one who will give a free ride to anyone. 'Not me!' she thinks, and pushes Dave's hand away, then mutters some excuse about her mother expecting her home. She runs off, looking over her shoulder until she turns the corner of the shed. Though Dave does not realise it, there never was a real chance that Susan would choose to let him have his wicked way with her. Many choices that she has already made over the course of a decade or more had already seen to that.

A few years later there is a lot of excitement in the family. Dad is going to California for a whole year to work on these new CAT scanners[3]. The rest of the family can go with him, though Susan will have to stay behind with Gran for a couple of months before she can leave, in order to finish off her A-levels. She is taking physics, chemistry and biology, and is a year ahead of most of her contemporaries. When she finally does get to California, it is wonderful. The rented house is so spacious, and only a bus ride from the beach. There are orange trees and flowers all over the place and even a swimming pool. Just what she needs after all the hard work that has gone into her exams. One evening, she and Dad go out to collect a takeaway and they bump into someone Dad introduces as Professor Libet. He asks Susan about herself, then says, 'Hey, maybe you could help me out with my tests. I'm always on the look-out for suitable subjects. There's nothing too difficult involved. They would just take up a bit of your time, and you would need to come down to the hospital for a day, maybe tomorrow. What we would ask you to do is…'.

Susan looks at him. He is quite short, but has great big, kind eyes. She had planned to go down to the beach tomorrow, but, what the hell, there would always be another day for that. 'Yes, I'll come', she says.

So there she is, sitting in the laboratory, wearing a sort of hairnet which is holding pads onto her scalp. Wires lead from the pads to the EEG machine, to record what is going on in her brain. Her own arm is resting on the arm of her chair, and she has agreed to bend her wrist at some moment of her own choosing. She is also keeping an eye on a big clock with just a single hand that is going round quite a bit faster than an ordinary second hand. Her mind drifts off, and she starts thinking about what she might be missing down at the beach. Then a sort of timer in her brain goes off and starts initiating the wrist movement. Around one third of a second later, her 'responsibility detector' notices what is happening and she feels a conscious wish to press the button. At the same time she sees, as instructed, that the clock hand is on the '8'.

When Professor Libet shows her the results at the end of the session, he is quite excited that her brain seemed to start to act well before she felt a wish to do so. 'But I just told it what to do', Susan says to herself. 'Of course I chose to bend my wrist, but that was yesterday. My brain was sort of on automatic in the lab. And I could have chosen differently if I'd wanted'. She does not say any of this out loud, for she is a polite girl. But she is quite right in one way. Her brain did indeed get to work before she became aware of it, but that doesn't show that consciousness played no part in causing it to get going – indeed quite the opposite. In fact, it suggests that unconscious brain activity may often occur at the behest of previous conscious choice. The difference between Professor Libet's interpretation and her own arose because Susan's awareness of a wish to press the button was not directly related to the conscious and unconscious mechanisms that actually caused the button press. Her 'responsibility detector' output was inevitably a little behind the times, because memory related like all conscious experience.

She may not have been quite so correct to say that she could have made a different choice yesterday. Maybe it was just possible. On the other hand, a lifetime of previous choices had turned her into the sort of person who was both interested in that sort of experiment and unlikely to refuse that sort of request for help[4]. Hannah, had she been there, might have been a good deal freer to choose whether or not to join the experiment, even though her own conscious choices compared to Susan's have had relatively little long-term influence on her life. Hannah has always been very much at the mercy of circumstance. Her mother was often drunk, and she had five different stepfathers in as many years. As a consequence, any conscious choices that she made were often quickly negated by events. She is less responsible for her character and circumstances than is Susan, but this very fact sometimes allows her a wider range of choice for things occurring over short timescales, of a few minutes, hours or days.

This is not to say that Hannah was bound to turn out as she did, nor Susan either. Sources of strength, or of weakness, can crop up in the most unlikely circumstances. And no one can know everything about what a child is experiencing, thinking or choosing – not even the child herself. There is the genetic lottery, too, which has some influence on outcomes. And some kids get brain damage from injuries, infections and so forth that can also affect how they turn out. All the same, the dice were always loaded against Hannah and in Susan's favour.

Y

Susan, of course, is a fictional character, selected to represent the silent majority of people who rarely make it into the tabloids or police records. The main purpose of telling her story is to show that free choice is not a one-second wonder; it relates to a wide range of timescales, from seconds to decades. It has its own pattern of growth and its own momentum. There are so many complex feedbacks occurring over these timescales that it can often be meaningless to ask whether some particular piece of unconscious neural activity is causing a conscious choice or is caused by one. Clearly, though, there is a sense in which consciousness has autonomy in that it can influence both its own future and the actions of its brain. The confusion that has arisen over these issues is mainly due to the fact that the actual causative pathways by which consciousness affects its brain are themselves mainly unconscious. This is not really surprising. When we choose to take a shower, for example, we have no idea of how our frontal lobes are issuing the relevant attentional commands, nor of how our motor centres get us to the shower. We simply decide to head for the bathroom and become aware that we are getting there, courtesy of mechanisms that, in detail, we know nothing about. As it turns out, though, the awareness that 'we' are are taking off our shirt, or whatever, is not a direct product of any of the conscious or unconscious mechanisms actually controlling our actions. It is associated with the output of a rather separate brain 'best-guesstimate module' that decides whether something occurring in the brain is due to the brain itself or to outside causes. We have been slow to appreciate this, and as a result it has caused a lot of conceptual muddle.

It's ironic, then, that what allows consciousness a degree of freedom from neural determinism (i.e. its pivotal position in the gateway to memory; its intimate involvement with memory) is also responsible for the most important constraint of all on an individual's choices. For it is not just faces and skills and character traits that Susan's choices have selected for learning or inclusion in her declarative memory. Also included are all sorts of concepts, prejudices, assumptions, habitual emotional evaluations and the like. These 'cognitive objects'[5] can often take on a life of their own. They can lead people in very different directions from those that they would freely have chosen as individuals uninfluenced by the quasi-independent behaviour of such objects. A degree of freedom from neural determinism is purchased only at the expense of considerable subjection to a sort of supra-individual, cultural determinism.

Some aspects of Susan's development can be seen as being like a story. For instance, she 'wrote' into herself her own view of what it is to be a good person, basing her account on a whole lot of experiences, including watching *Doctor Who* on television. The trouble is that the stories people incorporate can sometimes develop according to their own logic. To see what this means, and how it happens, we must leave neurology and psychology behind and look to literature and history.

Chapter 7

OBJECTS IN MIND

The term 'cognitive object' may sound rather cold and intellectual, but that impression is misleading. There's an ever-growing appreciation by neuroscientists of every persuasion nowadays that cognition is both founded on, and partly built with, emotion. The entities I'll be describing not only involve passions in their own make-up, they can raise passions of all sorts, too, which sometimes make it hard for people to grasp what they are like. We are generally very attached to the objects that we ourselves harbour, so can be reluctant to see them for what they are and may be poised to defend them against analysis and criticism.

Objects remote from us in time or culture, on the other hand, can be viewed with a certain amount of dispassion. It's best to start with one like this for practice. The idea of the 'Noble Roman' is suitable as it is quite well documented and had enormous influence from the beginnings of the Roman Republic right up to the earlier part of the 20th century, when it faded away due to the demise of most classical education. What components went in to making the idea, how did it survive, and what effects may it have had on people's lives?

Using an example from the past could cause problems opposite to those associated with living examples. Like a corpse undergoing dissection, which is inevitably lifeless and bloodless, a dead cognitive object can appear abstract and little more than a literary or historical conceit. So the Noble Roman seems to us today. But to a slave in a rich Italian household two thousand years ago the impression would have been quite different. The idea as well as the actuality of the Noble Roman were very real to such people and occupied much of their conscious lives day in and day out.

Luckily for my purposes, William Shakespeare brought the character to life in the person of Brutus, in *Julius Caesar.* After Brutus' death, his opponent Mark Antony eulogises:

> *This was the noblest Roman of them all:*
> *All the conspirators save only he*
> *Did that they did in envy of great Caesar;*
> *He only, in a general honest thought*
> *And common good to all, made one of them.*
> *His life was gentle; and the elements*
> *So mix'd in him that Nature might stand up*
> *And say to all the world, 'This was a man!'*

Let's take a look at the character and behavioural traits pictured here. Unselfishness of deed is the most obvious. There are hints at others, though Shakespeare does not go on to list the mixture of elements that made Brutus 'a man' (this is the penultimate speech in the play). The bit about 'his life was gentle' has to be taken with a pinch of salt from our perspective, since he was an assassin who had just fought an extremely bloody civil war and ended by committing suicide. 'Gentle' is an Elizabethan usage that invokes, not a Roman object, but a related, relatively modern English one: the gentleman. Maybe this is a hint that the Noble Roman did not die after all, but evolved into a somewhat different creature.

Earlier in the play, we are given more detailed information on Brutus's character. The unselfishness has a number of component ideas. Financial probity is one. Another is overriding concern about 'justice' for everyone, often at odds with personal inclination or affection. He was pretty stoical, too. His resignation in the face of news of the death of his wife, whom he loved, earned him the accolade: 'Even so great men great losses should endure'[1]. On the other hand, Brutus's nobility led him into foolish acts; for instance into allowing Mark Antony to make the funeral speech ('Friends, Romans, countrymen, lend me your ears' etc.) that switched popular support back to Caesar's party. It did not guarantee wisdom: Shakespeare hints that Brutus's strategic misjudgment, as well as the tactical pessimism of Cassius, lost them the crucial battle of Philippi. And he could be swayed into acting against inclination and loyalty, as when Cassius's manipulations resulted in his joining the assassins. He was, moreover, irritatingly smug. Here he is discussing the possibility of suicide:

No, Cassius, no: think not, thou noble Roman,
That ever Brutus will go bound to Rome;
He bears too great a mind...

Why were Brutus and Cassius both 'Noble'? Character traits alone cannot have been responsible. The underlying concept responsible is actually very simple, although it held an empire in thrall for four hundred years and continued to influence people for another thousand or more. A Noble Roman is a Roman noble, i.e. a Roman aristocrat. Brutus was especially noble because he was a descendant of another Brutus who, some 470 years previously, played a leading part in ejecting the Tarquin kings from Rome. The late 20th century historian Michael Grant has pointed out that Roman Nobles in the earliest days of the republic were simply the men of a small group of families called the patricians. Later on, their numbers were enlarged by including anyone who had had a consul or equivalent among his ancestors, patrician or not. All senators were noble but, as senators were limited to 300 (later 600) in number, not all nobles were senators. senators, especially, possessed great prestige, called *auctoritas*, which entitled them to respect and deference. The respect given to senatorial families long outlasted the Republic. Five hundred years after its death, when the Western empire too was dying, senators were still in a privileged and sheltered situation. Many of them in that epoch occupied vastly wealthy estates, which were islands of luxury and relative order amid ever-increasing general poverty and social chaos. Moreover, the idea that respect was still owed to Roman nobles in this age of disintegration remained very much alive. Says Grant in his *History of Rome*, 'when… Claudian of Alexandria, the leading poet of the age, wants to write flatteringly about his own contemporaries, he can never stop comparing them to the Scipios and Catos and Brutuses…'.

Cicero, a famous orator (but not himself noble) and an older contemporary of our Brutus, showed how uncritical was the respect felt for nobles. When defending a client named Gaius Rabirius from a trumped-up charge of murder – trumped up, as it happened, at the instigation of Julius Caesar – he thought it a valid debating point to simply list the names of the aristocratic men who were on Rabirius's side. They surely, was the implication, could not possibly be involved in any wrong-doing. The nobles themselves, or many of them, seem to have shared the general view. Respect for their own ancestry was very important to them, and a desire to emulate famous ancestors was a constant motivating force.

The Roman historian Livy, writing quite soon after the death of the Republic, also had a highly developed notion of the moral standards and general worthiness of nobles exemplified by antique heros and the current emperor, Augustus Caesar (Julius's successor). Subsequently, the idea that their characters were anything to write home about seems to have suffered a temporary eclipse. Suetonius's *The Twelve Caesars* was written about 150 years after the assassination. The book has never been out of papyrus, parchment or print since it first appeared. But the nobility of the Noble Roman is conspicuous by its absence from this work. There is a reference to an unfortunate historian (unnamed) who was executed by Tiberius, around 70 years after Julius's death, for referring to Brutus and Cassius as 'the last of the Romans', so suggesting that the 'justice' concept was still active then. Suetonius also reports some doggerel written on a statue of Caesar, which appears to indicate that emulating ancestors was still considered important. But that's all. The overall impression given is that nobles were a shifty bunch of corrupt, arrogant and self-seeking hypocrites. Suetonius, however, was something of a tabloid journalist before his time. The greatest historian alive at that time, Tacitus, seems to have retained the concept, but it was not nearly as central in his work as in that of Livy a hundred years previously.

A weakening of the concept in Suetonius' time is entirely understandable in a people who had just experienced 50 years of rule by vicious, crazed or incompetent aristocrats such as Tiberius, Caligula and Nero. The surprise is that it should ever have reassembled. One might have expected that senatorial rank and the idea of admirable character would have remained strangers, but this was not so. The process of reassembly may have been helped by a subsequent discovery that non-aristocratic emperors could be just as bad as any Nero. Moreover the concept seems always to have been amazingly resilient in Roman society, having in Republican times survived the destruction of whole armies by incompetent consuls.

The emperor Marcus Aurelius provided another restorative influence, even though he was not himself from old nobility (he had been adopted into it). His *Meditations*, recorded about 60 years after Tacitus's *Histories*, was one of the most popular books ever written. It's an account of his experience and musings as a sensitive man, doomed to struggle with uncongenial military and administrative tasks, who nevertheless always puts duty first and copes with a gloomy but serene stoicism. It provokes admiration and expresses almost all that was best in Roman governors. Marcus Aurelius was perhaps responsible for aspects of Shakespeare's portrayal of Brutus, particularly that nobility of soul was at least as important as nobility of ancestry – a notion that is very unlikely to have occurred to Brutus himself when he was alive.

Despite some changes of emphasis, it is remarkable how closely Shakespeare's portrait of Brutus corresponds to what is likely to have occupied the minds of real Noble Romans of the time. They did value financial probity – as demonstrated by the frequent scandals over corrupt practices by provincial governors. They certainly felt that 'justice' (meaning mainly freedom from coercion) should apply to them and their equals, if not to lesser citizens. And they were usually amazingly stoical. The unselfishness in Shakespeare's account was probably, in the case of the real Brutus, based on a firm idea that family, peers and state are more important than self. Only a hundred years before Caesar's death, the aristocratic Cato the Elder had refused to include the names of any Roman generals in his histories, on the grounds that individualism should be discouraged.

Suicide, of course, was something that Romans contemplated without Shakespeare's moral ambivalence. From a Roman's point of view, to end one's own life was greatly preferable to execution because it was less undignified and because property could usually pass to one's heirs instead of being confiscated. It could be laudable even in circumstances that might appear to us to warrant a quite different response. Tacitus, for instance, admiringly recounts that a centurion whose report was disbelieved by the usurper emperor Vitellius killed himself with the words:

> Well, since you need overwhelming proof and have no further use for me whether alive or dead, I will supply you with evidence you must believe. (Translated by Kenneth Wellesley)

Some 1600 years separated Brutus and Shakespeare, yet the idea of the Noble Roman changed little in all that time. It gained a few components and lost others, but remained recognisably the same, despite having fallen apart or gone into retreat for a while when Tacitus and Suetonius were active. There was a much more serious eclipse later on. If Marcus Aurelius's *Meditations* was one of the most influential books ever written, it was nevertheless outshone by a work published some 200 years later –

the *Confessions* of St Augustine. The Noble Roman gets just one look in here in a brief passage about Victorinus, 'an old man of great learning' who had been tutor to several senators and awarded the 'honour' of a statue in the Forum. St Augustine hadn't much time for consular ancestry or the like, it seems.

For another thousand years, the *auctoritas* of the Noble Roman was usurped by quite different figures: desert fathers, monastic organisers, chivalrous knights, feudal kings and so on. Writing about 200 years before Shakespeare, Chaucer seems to have no inkling of the concept in his *Canterbury Tales*, although he probably had access to much the same historical sources as did Shakespeare. Of course the Renaissance, with its revival of interest in all things classical, had happened in the interim, which may account for the completely different emphases given by the two writers. Here is Chaucer's account of Caesar's death:

Up to the Capitol this Julius went
A certain day as he was wont to do.
There he was taken by the malcontent
False-hearted Brutus and his scheming crew,
They stabbed him there with daggers through and through,
Many the wounds, and there they let him die.
(modernised by Nevill Coghill)

Renaissance or no, it does seem odd that the Noble Roman should have come to life again after so many centuries. It was not only in Shakespeare's play that it did so. The title 'senator' was a popular choice for senior legislators in a whole range of republics established after Shakespeare's time. But the original core components of the concept were the importance of consular ancestry and the *auctoritas* entailed by it, neither of which had the slightest relevance in Shakespeare's day, let alone subsequently, when even the Eastern Roman Empire (Byzantium) was defunct, having been conquered by the Ottomans in the previous century. Nevertheless, the concept did revive, and perhaps achieved its most faithful replication or resurrection in the form of a man named George Nathaniel Curzon, Viceroy of India from 1898 to 1905.

All his life, he was plagued by a lampoon written when he was a student at Oxford:

My name is George Nathaniel Curzon,
I am a most superior person,
My cheek is pink, my hair is sleek,
I dine at Blenheim once a week.

(Blenheim is the palace of the Dukes of Marlborough.)

Despite his many complaints about this verse, the epitaph he wrote for himself, found among his papers after his death at the age of 66, suggests that he fundamentally agreed with it:

In divers offices and in many lands
As Explorer, Writer, Administrator and Ruler of men,
He sought to serve his country
And add honour to an ancient name.

Clearly he shared Brutus' preoccupation with ancestry and the '*auctoritas*' so entailed. He was also financially scrupulous, despite being often short of money in his earlier life. For instance he opposed a motion that Members of Parliament, such as he was at the time, should be paid a salary, lest the job should attract 'the idle, the necessitous and the unscrupulous'. (Impractical though it was, one might at times wish that his opinion had been heeded!) Of other important ideas, he possessed a high degree of stoicism, especially in relation to his chronic back pain which made a nightmare of many activities and official duties. His views on suicide are unknown, though he was certainly physically brave in the Roman manner. The smugness bit was there, too. 'Oh that to every Englishman in this country, as he ends his work, might be truthfully applied the phrase "Thou hast loved righteousness and hated iniquity"', he said in his farewell speech to India. An implication, of course, is that the phrase did apply to himself.

The similarities are so great that it is tempting to wonder whether Brutus himself may not also have shared some of Curzon's other characteristics; his intellectual adroitness, for instance, his love of orotund phraseology and his capacity for painstaking hard work. Maybe Brutus too had the rather elephantine, but lively, humour that Curzon showed before middle-age and misfortune made him chronically tetchy. Curzon, for instance, genuinely enjoyed the following 'tribute' offered to him by a friend after a brief spell in a post at the India Office in London, which he occupied some time before being made Governor:

> I apprehend that... [Curzon] has done more from sheer gaiety of spirit to destroy our empire in Asia than any average Viceroy could effect in ten years or the Emperor of Russia in a hundred. This is a record of which any man may be proud.

The coincidences between these two characters, separated from one another by nearly two millennia, have much to do with Curzon's excellent classical education and the related ideas about aristocracy that developed in Republican Rome and 18th century England. Moreover, Curzon is reported to have been most impressed by a reading of *Julius Caesar*, the play, which he attended at the age of ten. Other parallels between the two characters, or at least between Curzon and Shakepeare's portrayal of Brutus, are likely due to chance[2]. For instance, Curzon, too, genuinely loved his first wife, who died before him. He, like Brutus, had a circle of cronies who were constantly anticipating great deeds from him. Though not driven to suicide, Curzon was driven into the political wilderness by a complex character, Lord Kitchener, reminiscent of Mark Antony's principal ally Octavian at his most sinister.

Y

This impressionistic account of representations of an enduring cognitive object highlights aspects of its constitution and behaviour that may apply more generally to similar objects. For instance, it seems that they may contain a few core concepts surrounded by a cloud of associated ones that can change without loss of the object's identity. Maybe these objects can contain no more than around seven concepts at any one time, because of the limited capacity of working memory (see Chapter 3). Nobility of lineage and character, love of freedom, physical stoicism and financial honesty are all essential to the Noble Roman, though the relative importance of these notions varies from representation to representation. Ideas that modify concepts of what constitutes freedom or nobility of character, say, may differ in different representations. Other subsidiary notions, such the validity of suicide, can be absent altogether without obvious harm to the object.

But when the core concepts become separated from one another the object disappears and its seeds lie dormant in books or other records, or else they get incorporated in quite different representations. The idea of noble lineage, for instance, was subsidiary in a very different character from the Noble Roman, the Knight Errant, one of several that helped span the centuries between successive periods in which the Roman appeared. There is room for endless debate over whether it was the survival of some of Livy's works, along with other similar books, or the continuing life given to important concepts in other cognitive objects, that allowed resurrection of the Noble Roman a thousand years after the Roman Empire had died. Both circumstances probably helped and one could argue about their relative importance for a long time without reaching any conclusions.

The assembly or reassembly of an object like the Noble Roman seems to depend on cultural and social circumstances at the time – the general pool of ideas active in a given time and place – perhaps aided in this case as far as reassembly is concerned by the capricious genius of Shakespeare.

Another important feature suggested by the Noble Roman is that objects can be 'incarnated', so to speak, in a weak sense or a strong sense. The weak sense would be exemplified by someone reading or seeing *Julius Caesar* and having their awareness and imagination temporarily occupied by the character of Brutus. The strong sense is exemplified by Curzon who, at least for part of his life, essentially *was* a Noble Roman. In fact there's probably no clear dividing line between strong and weak incarnations. And we are not discussing abstract ideas or transient acts of imagination only. Real people can, in a sense, *be* objects of this sort. What does all this imply for the individual free will of the characters concerned? We need to look at some more examples before reaching any tentative conclusions.

Chapter 8

MORE COGNITIVE OBJECTS

The Noble Roman was an idea, or ideal, of a type of person. Other ideas are similar in their make-up and behaviour but are not personal. They have to do with the forces that affect how groups of people act, rather than with how individuals behave. Lewis Mumford, the distinguished 20th century American social philosopher, held an entity of this sort – he used the word 'archetype' – responsible for the Greek and some of the later Utopias. He suggested that relevant concepts coalesced at the time of the very first Neolithic cities and allowed the development of 'human machines'. These were groups of people, subordinated to a common purpose, organised into building pyramids, ziggurats or whatever. The cognition subsequently lived on, was responsible for the repellently regimented nature of most Utopias (e.g. Plato's *Republic*) and finally paved the way for the true machine age whose growth accelerated so dramatically in the Industrial Revolution. He claimed, in other words, that archetypes can persist for millennia and dramatically affect the behaviour of people in cultures that harbour them.

There is a temptation to try to fill in the outline Mumford sketched, but a problem, living as we do in a time pervaded by the Utopian idea, is that many of its component concepts appear to be self-evident truths. There are easier examples that illustrate the necessary points just as well. The two described here are fairly well documented and appear to us outlandish even though they are still very much alive. They are simple compared to, say, the Philosopher's Stone which might otherwise have been a candidate. One is the dancing manias, the other the millennial movements; both flourished in the mediaeval period.

As usual, there is a caveat to issue before getting down to the nitty-gritty. The dancing manias in particular have often been mistaken for physical illnesses by both contemporaries and moderns. Infection or poisoning of some type has usually been blamed. Ergot poisoning from badly stored cereals is a favourite modern explanation, for example. In fact, if they were illness-like at all, it is not physical illness that they resemble. They are far more akin to a craze like the ones for hula hoops or texting, or to the strange 'neurological' epidemic that afflicted managers throughout the Anglophone world in the 1980s. These people suddenly lost the ability to utter the words 'now' or 'at present', and were compelled instead to mouth gobbledygook at every opportunity: 'at this moment in time'. Luckily they are now recovering (though a strong urge to say 'this is soooo not me' shows some signs of taking over). The phenomena I shall describe differ from what happened to these people only in degree, not in kind. Both dance manias and millennial movements have also sometimes been blamed on diabolic causes. I subscribe to the diabolic theories no more than to physical ones – despite a natural urge to do so in the case of one or two managers that I've met!

Dancing manias

Two main dancing epidemics broke out in the Middle Ages. One, 'St Vitus' dance' or 'St John's dance', occurred mainly in Germany and surrounding areas. The other, initially confined to Apulia in Southern Italy, was named 'Tarantism'. The term 'St Vitus' dance' has also been applied to motor disorders, such as rheumatic chorea, but these physical illnesses are definitely not our concern here[1]. It would be hard to improve on Justus Hecker's[2] lively and enthralling account of the dancing manias, written in 1837 when he was professor of medicine in Berlin:

> ...the attack commenced with epileptic convulsions. Those affected fell to the ground senseless, panting and laboring

> for breath. They foamed at the mouth, and suddenly springing up began their dance amid strange contortions.

Tarantism tended to be a generally more elegant affair than St Vitus' dance. Special tunes called Tarantellas were composed to accompany it and witnesses were sometimes surprised that afflicted peasants should move with such apparent skill. Sufferers could dance for hours or days and occasionally died of exhaustion. Attacks were sometimes preceded by sensations of oppression or of pain, and hallucinations could occur during them.

Dance manias were usually group happenings, though sporadic individual cases were recorded. All ages, from childhood on, and both sexes were affected. Aggressiveness and lewd behaviour sometimes accompanied attacks; occasionally episodes of dancing would degenerate into riots or orgies. On the other hand, women expecting an attack were often noted to sew up their skirts into culottes to prevent involuntary exposure later on – knicker-wearing did not become widespread till the 19th century, and even then was mostly confined to the middle classes.

Dancers were generally very sensitive to music, which could provoke attacks but was more usually observed to have a benign effect on those already afflicted. The provision of bands of musicians to aid groups of dancers, an expensive undertaking, was known to be an appropriate treatment. People with St Vitus' dance were often driven to extra frenzies by a sight of the colour red. Indeed, some local authorities forbade the wearing of red clothing because of this. Those with Tarantism generally lacked this specific aversion, even though individuals occasionally showed sensitivity to colours of one shade or another.

Outbreaks of St Vitus' dance occurred in 1027, 1237 (when upward of a hundred children were said to have been suddenly seized at Erfurt, and to have proceeded dancing and jumping along the road to Arnstadt) and 1278, followed by major epidemics in 1374 and 1418, which involved many areas in Germany, the Low Countries and northern France. More than 500 people were said to have been affected in Cologne during the 1374 epidemic and as many as 1100 in Metz. The condition was still seen in the 16th century, but 'grew every year more rare, so that, at the beginning of the 17th century, it was observed only occasionally in its original form', wrote Hecker.

Tarantism followed a more endemic course. The earliest written account dates from the mid-15th century, but it had probably been established in Apulia for more than a hundred years before this. Probably not by coincidence, the condition began to spread to other parts of Italy fairly soon after having been described in print, becoming most prevalent at around the time that St Vitus' dance was fading out in its own homelands.

The core concepts responsible for these strange phenomena are not entirely clear. There are probably elements of the same motivations that produced bacchic revels in the classical world; that is, ideas of release and celebration through dance. Hecker thought that the alternative name for St Vitus' dance, St John's dance, was significant, as St John's day was a pagan festival that had, like Christmas, been Christianised. In Germany a traditional St John's day activity was to leap through a special bonfire, the 'Nodfyr', to prevent disease for the following year. A core concept, in other words, may well have been a notion that leaping or dancing protects from or cures illness – physical, mental or spiritual. Another could have been the identification of dance with worship, as exemplified by King David dancing before the Ark of the Covenant. Subsidiary notions that modified the course of the affliction from time to time included ideas that dancers were not responsible for violent or promiscuous behaviour while afflicted and that (in the case of St Vitus' dance) the condition might have diabolical causes.

The differences between the two principal forms of Dance were due to a small group of clear-cut concepts; namely that Tarantism was the only cure for the otherwise fatal bite of what was then called the 'tarantula' spider, which originally lived only in Apulia. There are three related ideas here: the means of cure, the consequences of being bitten and the alleged habitat of the spider. This spider was not what we mean nowadays by a tarantula; it was small, inconspicuous and harmless. Thus any minor insect bite might be due to it and require urgent dancing to prevent terrible consequences, so it was supposed. The history of the toxicity belief may date back to the Romans, who thought that a particular sort of lizard, actually innocuous, was severely poisonous. The Latin name for this lizard was *stellio*, but apparently it became it became known as a *terrantola* to early Italian speakers. In Apulia the name, along with the idea of toxicity, seems to have transferred over time to the arachnid world – maybe because a few mildly poisonous spiders do live there – and became joined, somehow, to the additional notion that the poison could be sweated out in dance.

However harmless the spider, the idea was not innocuous. Says Hecker:

> Quinzato, Bishop of Foligno, having allowed himself by way of a joke to be bitten by a tarantula, could obtain a cure in no other way than by being, through the influence of the tarantella, compelled to dance. Others among the clergy, who wished to shut their ears against music, because they considered dancing derogatory to their station, fell into a dangerous state of illness by thus delaying the crisis of the malady and were obliged at last to save themselves from a miserable death by submitting to the unwelcome but sole means of cure.

Presumably the bishop was a rationalist who wanted to show the ignorant peasants that their beliefs were ill-founded. But the joke was on him. Individual free will, it seems, is no match for this sort of idea when it is in full flow.

Hecker took the very reasonable view that dancing manias spread through imitation, thus anticipating by 140 years Richard Dawkins' description of how his 'memes' (supposed 'units' of culture analogous to genes as units of heredity) propagate[3]. More often, Hecker used 'sympathy' or 'morbid sympathy' as synonyms for 'imitation', which tend to emphasise the power and involuntary nature of the process:

> The whole world is full of examples of this afflicting state of turmoil, which, when the mind is carried away by the force of a sensual impression that destroys its freedom, is irresistibly propagated by imitation. Those who are thus infected do not spare even their own lives, but, as a hunted flock of sheep will follow their leader and rush over a precipice, so will whole hosts of enthusiasts, deluded by their infatuation, hurry on to a self-inflicted death.

It's understandable that some times and places within the mediaeval period, when there was an unhappy combination of religious enthusiasm, plague, famine or war, should have provided especially fertile soil for the dancing manias. But an identical condition was common in Ethiopia at least up to the mid-19th century, where it was also associated with St John. Scattered outbreaks of similar phenomena, some with more emphasis on convulsions than in earlier cases of St Vitus' dance, occurred in France, Lancashire, Cornwall, Scotland and the Shetland Islands in the 18th century. Some were short-lived, but others became institutionalised and endured for half a century or more. For example the sect of the 'Jumpers', a Quaker-like group of religious enthusiasts, which originated from Cornwall, or 18th century cases in Scotland, of whom it was reported in the *Edinburgh Medical and Surgical Journal*, 'when the fit of dancing leaping or running comes on, nothing tends so much to abate the violence of the disease, as allowing them free scope to exercise themselves, till nature be exhausted.... In some families it seems to be hereditary: and I have heard of one in which a horse was always kept ready saddled, to follow the young ladies belonging to it, when they were seized with a fit of running'.

In the 19th century, there are records of brief outbreaks and sporadic cases in the USA and in Japan, which had only recently reopened its ports to foreigners. The 1867 outbreak in Japan seems to have been very like a European dancing mania. But a possibly related condition appeared in 1771, when Japan was still isolated: groups of people went about carrying placards, some absurd or obscene, shouting, singing and getting ever more excited. It might be that, for typical St Vitus' dance to occur, there must be some Christian tradition. Ethiopia was always Christian, while Japan experienced some revival of interest in the religion during its 19th century political upheavals. The Muslim whirling dervishes cannot be used to counter this proposal since they represent a less involuntary phenomenon that has an overtly religious goal, and thus probably shares few of the concepts of the European dancers. However, an epidemic of classical St Vitus' dance broke out in Madagascar in 1863 as part of a movement intended specifically to oust Christianity and restore ancestor worship. Any influence there of Christian ideas cannot have been due to conscious acceptance of them, but perhaps they had, nevertheless, some involuntary effect.

The most remarkable thing about involuntary Dance as cognitive object is that it could suck people in willy-nilly, determining their behaviour and the focus of their awareness over quite long periods of time. It provides a particularly striking example of the power of groups to replace individual decision-making with a 'group consciousness'. Biological mechanisms can be proposed to account for these features – a hard-wired herd instinct, for instance, or, more plausibly, a physiological tendency for rhythmic movement to reduce the ill effects of stressful circumstances or to induce auto-intoxication. But there is also the question of why the movements in question took such a stereotyped form in so many different times and places. It looks as though the objects themselves may have a dynamic of their own, perhaps contributed to, but not wholly caused by, our biological make-up.

Millennial movements

The superb account that historian Norman Cohn gave of these terrible phenomena in his 1957 book *The Pursuit of the Millennium* should be included in the National Curriculum as an object lesson in the follies of belief. Millennial ideas of course got a boost recently with the arrival of 2000 or, for those who wanted to celebrate the 'real' millennium, 2001. In fact, though, the years 1260, 1421 and 1534 were regarded as particularly significant in various previous outbreaks of millenarianism. In most traditions 'millennium' refers to the expected duration of the whole thing, as in Hitler's 'thousand year Reich', not to the starting date.

The millennial idea is more complicated in operation than those described hitherto, as it depends on various subsidiary entities in order to exert its full effects. These include the Prophet, the Last Good Emperor, Antichrist and the Messiah. Not all were present in every movement, but one or more was always needed to unleash the Millennium. Cohn's helpful list of core concepts of the object runs as follows:

> Millenarian sects or movements always picture salvation as:
>
> (a) collective, in the sense that it is to be enjoyed by the faithful as a collectivity;
> (b) terrestrial, in the sense that it is to be realised on this earth and not in some other-worldly heaven;
> (c) imminent, in the sense that it is to come both soon and suddenly;
> (d) total, in the sense that it is utterly to transform life on earth, so that the new dispensation will be no mere improvement on the present but perfection itself;
> (e) miraculous, in the sense that it is to be accomplished by, or with the help of, supernatural agencies.

Unfortunately, the concept was based on an apocalyptic, and often apocryphal, literature which also included more obviously malignant ideas, especially that the advent of the Millennium would be preceded by the reign of Antichrist and that there would be a need for the faithful to cleanse the world of his servants before salvation could be accomplished. These ideas surfaced so often that they could be added to the list of core concepts were it not for their apparent absence from a very few manifestations of the object. Occasionally conversion rather than annihilation of the unfaithful was the goal, but the Christian apologist Lactantius[4], writing in the fourth century, expressed a more typical attitude:

> But that madman [Antichrist], raging with implacable anger, will lead an army and besiege the mountain where the righteous have taken refuge. And when they see themselves besieged, they will call loudly to God for help, and God shall hear them, and send them a liberator. Then the heavens shall be opened in a tempest, and Christ shall descend with great power; and a fiery brightness shall go before him, and a countless host of angels; and all the multitude of the godless shall be annihilated, and torrents of blood shall flow...

Later, of course, the righteous came to feel that they should help Christ and the angels with the work of annihilation, or even undertake it themselves if Christ appeared tardy about getting on with it.

Thanks to St Augustine and others, some of the core concepts of the idea – terrestriality, imminence and totality – were suppressed among educated classes for a long time, but continued to rumble on, says Cohn, 'in the obscure underworld of popular religion'. These concepts erupted at around the time of the first crusade. The Pope who launched this allegedly did so with the sensible and limited aim of supporting a tottering Byzantine empire in its battles against encroaching Turks. The principal propagandist for the second crusade was St Bernard, who may have had more extensive goals in mind. Both the Pope and

St Bernard were, in the event, overwhelmed by a wave of millenarial enthusiasm which, almost incredibly, led to the capture of Jerusalem and the establishment of Latin states in the near East. One of these, the county of Edessa, was quickly extinguished, but the others endured, in ever-dwindling form, for nearly as long as the British Empire in India. It also led to enormous suffering at an individual level and, in due course, to the great weakening of the Byzantine empire. As so often happens, the Pope had achieved the opposite effect to that originally intended.

That Jerusalem was captured, instead of Byzantium aided, was largely due to a subsidiary concept. People have always had problems in picturing the nature of the total change that is to overtake society according to the Millennium idea. There was a persistent current of thought that post-millennial life would be based on 'a new Jerusalem', whatever that might mean, so it made obvious sense to get one's hands on the old one first. Some have assumed that the incentive to capture the city came from plenary indulgences offered to crusaders or from a desire to visit the holy places. Neither of these explanations suffices, as the indulgences would have applied wherever fighting against Muslims had taken place, while, prior to the first crusade, Muslim authorities had generally been helpful to Christian pilgrims. It was a specific desire to possess the site of the coming revolution that was responsible for its capture. Crusaders themselves often referred to the 'wickedness' of the Muslims in destroying the church of the Holy Sepulchre in Jerusalem – something that had been ordered by an unstable Egyptian caliph and had shocked Christendom. However, this was war propaganda only, as the event happened 88 years before any crusader left France and the church had in any case been rebuilt (by a Byzantine emperor) about 30 years after its destruction.

After Jerusalem was lost and interest in the classical world revived, linked with the Renaissance, a more detailed picture of the outcome of advent of the Millennium was constructed on the basis of Greek myths of what life was like in the 'golden age'. Among its features was community of property, which was sometimes regarded as including spouses. The consequences of this were at least as bizarre, if not quite as bloody, as the unexpected capture of Jerusalem (among the many deaths directly or indirectly due to crusading were almost the entire Muslim population of Jerusalem, numbering several thousands).

Millennial enthusiasm for crusading was a mainly lower-class phenomenon. The upper classes who were involved usually had quite different agendas, such as to set themselves up as independent rulers, to carve out an estate, or to carry out the Pope's policy, though some got carried away by the general spirit. When not organised by cooler heads, the consequences of crusading were often disastrous for participants as well as many of those whom they encountered along the way. In the so-called children's crusade, thousands of children from France and the Rhine valley headed for the Mediterranean, which was to dry up before them and allow passage to the Holy Land. Nearly all drowned, starved to death or were sold into slavery. The Shepherd's crusade of 1251 resulted from the activities of two preachers who raised an army of young enthusiasts, some of whom ran to join without saying goodbye to their kin or making provision for the journey. They soon degenerated into a marauding horde; most died or were killed before they could leave Europe. These are just two striking examples of many similar movements.

A problem for these bands of enthusiasts was the urgent need to take action against servants of Antichrist. Muslims, a legitimate target in the eyes of the church, were often inaccessible, but fortunately, in their view, others were to be found nearer to hand. Jews (because their race had 'killed Christ'), priests (because many conspicuously failed to live up to ideals of poverty and chastity) and the rich (for obvious reasons) were convenient targets. Crusading hordes massacred Jewish communities in Mainz, Cologne, Metz (which is now in France, but wasn't then) and in parts of France too, among many other places, before ever leaving their own countries. The efforts of church and secular authorities to prevent massacres were usually unsuccessful. Between 4,000 and 8,000 Jews were killed in 1096 alone.

The necessity of death for servants of Antichrist was not the only idea commonly joined to core concepts of the Millennium. Another was that the righteous would be prepared for the great day by undergoing a time of troubles and distress, which would purify them. Flagellants seized on this and took matters into their own hands. In the same spirit that inspired so many millenarians to help Christ destroy the wicked, these people set about purging themselves. They lashed themselves with whips, sometimes tipped with hooks so that chunks of flesh were torn out, for hours on end. Large groups of self-whipping men and boys first appeared in Italy in the year 1260, 'with such rapidity that to contemporaries it appeared a sudden epidemic of remorse'[5].

This first epidemic faded out in Italy after a year or two, but crossed the Alps to surface in Germany in a more organised form. Flagellants there possessed rituals, songs and even a sort of uniform. Over the succeeding two centuries, groups could appear anywhere in Europe, particularly when times were hard; there was an especially severe outbreak in 1348–49, apparently precipitated by the Black Death. Eventually the church decided to suppress the movement, at least partly because some of its leaders developed Messianic or 'Last Good Emperor' pretensions. The Inquisition swung into action, and the final group of flagellants was discovered and burnt at the stake around 1480 – in central Germany, which had been an ancient heartland of the movement.

The strange consequences of eliding Greek ideas about the golden age with apocalyptic ones were most clearly demonstrated by events in the town of Munster in the year 1534. During the aftermath of all the upheaval caused by the Reformation this town was taken over by a millenarian sect called the Anabaptists who forcibly converted or ejected all citizens not of the same mind as themselves. They soon found themselves besieged by rather unenthusiastic armies raised by the prince-bishop of Munster who, luckily for himself, was not a resident of the town.

The first sectarian leader in the town, 'a tall, gaunt figure with a long, black beard', was quite moderate. He merely sanctioned the requisitioning of all private property, the ejection from the town (in February) of everyone, including the sick and the pregnant, who disagreed with him and, interestingly enough, the burning of all books apart from the Bible. He was soon dead. It was revealed to him that if he made a sortie with only a few followers he would rout the besiegers, but this turned out not to be the case.

The next leader, a man named Bockelson, was more colourful. One of his first acts was to run naked through the town in a frenzy after which he refused to speak for three days. When he broke his silence, it was to command an absolutist reorganisation of the council and laws. The regulation of sexual matters was particularly confused. At first there was a severely puritanical attitude to them, but later Bockelson, in common with many previous millenarians of the same ilk, discovered that this attitude was wrong. His predecessors had often simply advocated free love for all, community of sexual possession along with community of property. Bockelson, however, modified this ideal to fit local conditions by introducing compulsory polygamy. Fifty objectors to his policy were executed straight away and a number of women who refused to comply were similarly dealt with over succeeding months.

There was actually an opportunity to break the siege in August when the prince-bishop's mercenaries had temporarily wandered off. But, instead of taking it, the great leader was fully occupied in having himself crowned not only as Emperor of the Last Days but also as Messiah. He spent some of the brief time remaining to him, before the besiegers won, in personally slaying wrongdoers who came to his notice.

Cohn makes the point that millenarianism flourished among rootless lower classes, particularly those who had got sucked into mediaeval precursors of industrialism such as the weaving factories. It was generally exacerbated by the social upheavals of war, plague, poor harvests and the like. Leaders, the would-be prophets and messiahs, were often not from the same class. They tended to be clerics of the perpetual student variety, or sometimes minor nobility[6]. Although their actions appear to us bizarre, they were mostly neither mad nor insincere. They were simply overwhelmed by the idea of the Millennium. Even Bockelson[7], faced with inevitable torture and death, declared with probable sincerity that he had always sought God's glory. Followers of these sects frequently recanted when the sect was suppressed, but leaders would usually go to the fire affirming their own truthfulness or, in some cases, with laughter.

Lest it be thought that only the uneducated or half-educated can nurture disorders of this sort, Cohn in a further book on witchcraft and related topics describes a condition affecting mainly the well educated. This was the idea, which he convincingly argues was totally false, that hidden sects exist which seal their secrecy with fearful oaths involving eating human flesh or drinking human blood, most often obtained from babies. The notion first surfaced in Rome in connection with a political conspiracy (the Catiline affair which Cicero played a leading part in suppressing). Then it flourished in relation to Roman fantasies about the early Christians. After going underground for a time, it resurfaced to infect the Church's dealings with a range of heretics of the Middle Ages (Cathars, Waldensians etc.), aiding their persecution and slaughter. Then it got attached to the witch-hunting manias of the 15th to 17th centuries, and played an essential part in them.

It reappeared not too long ago in only slightly modified form in Orkney, in connection with alleged Satanic abuse of children by a group of islanders led by a minister of the church. The minister was believed to have danced vigorously and nudely, often in freezing weather, throughout most of the night during these horrific ceremonies. And Orkney nights are indeed long in the winter. Later, following a judicial enquiry and the return of reason, it turned out that he was over sixty and suffering from a heart condition at the time of these supposed exertions. In the meantime, social workers backed up by police had descended from ferries and taken lots of children into care to protect them. There is no reassurance to be gained from the fact that the notion apparently no longer infects priests or the judiciary, but has transferred to social workers. Something with such a long history of colonising the seats of power may be a good indicator of which group in a society really is in charge!

Individual accusations of witchcraft, in particular, were usually a family or neighbourhood matter, precipitated by ill feeling, just as murder is commonest in the family. Interestingly enough, a frequent 17th century source of antagonism leading to such accusations was quarrels between neighbours over alms-giving. But the climate in which accusations could flourish was, and is, an involuntary creation of the best-educated people, who might be imagined to possess more individual autonomy than most. As individuals, most of them were probably well-intentioned, but they were nevertheless driven by the stories current in their societies into beliefs and choices with consequences that were very far from benign.

The Millennium itself, of course, still surfaces only too frequently in sects like the Branch Davidians, whose enclave was besieged by the FBI, or the people (the 'Heaven's Gate' group) who believed that a spacecraft lurking behind comet Hale–Bopp was coming to fetch them to a better world. And most participants in bizarre events like these appear to have been ordinary, regular folk, very like you or me, both before getting caught up in them and indeed afterwards – if they survived. Mass suicide seems to be a more usual outcome of such aberrations nowadays than mass murder of outsiders, at least in the Western world. But that's only a small comfort.

Y

The Dance and the Millennium, like the Noble Roman, have a group of core concepts surrounded by a cloud of more mutable ones, which determined details of how the ideas manifested themselves. Perhaps this is characteristic of all such impersonal objects. Being non-personal, they could not 'incarnate' themselves in quite the same way as the Noble Roman, but were certainly able sometimes to overwhelm personality and make people act in uncharacteristic ways. Moreover personal concepts associated with the Millennium, the Messiah for instance, were only too often incarnated.

Both objects were contagious. St Vitus' dance and Millenarianism were epidemic, spreading much like the bubonic plague which so often appeared to trigger attacks of both. Tarantism was endemic, like chickenpox. A cursory look at their component concepts shows why this was so. The core ideas of the Dance (prevention of ills, worship etc.) and the Millennium (imminence, miraculousness etc.) are vague, capable of many interpretations and attributable as relevant in many different situations, albeit usually adverse ones. Therefore, any time of troubles left society vulnerable. The Millennium was particularly virulent because one of its associated concepts is that a time of troubles before its advent would purify the righteous. One can certainly sympathise with people who interpreted the Black Death in this light. It killed one third of Europe's population at its onset in the mid-14th century.

Tarantism behaved differently because of the three well-defined ideas added, in its case, to the core of the Dance – that an Apulian spider bite would prove fatal if not treated by immediate dancing. These provided a predisposition for the Dance to appear whenever anyone living in a particular geographical area had an attack of malaise or anxiety that might conceivably be associated with a spider bite. There's no surprise in the fact that the condition was endemic for several centuries, though it is perhaps odd that many people remained immune.

There is a hint that other notions that are in the air, such as that of a time of troubles being due to God's purifying wrath, can greatly aid the spread of some cognitive objects. The history of the Witch, mentioned in passing, provides an especially concrete example. It is obvious to most of us nowadays that almost anyone under torture will say whatever they think an inquisitor wants to hear. Whatever the case in relation to contemporary inquisitors, the likely truth-distorting effects of torture were certainly not obvious to Church inquisitors or judicial inquirers of the late Middle Ages, few of whom were sadistic cynics. They were conscientious people on the whole, with a genuine wish to save the souls of heretics or to apply the law justly. They believed the stories of cannibalism, commerce with the devil and so on that poor women so readily spilled out when put to the question. A few of the women may themselves have genuinely believed that they had taken part in the gruesome and marvellous acts attributed to them by inquisitors, such is the power of ideas to mould fantasy, dream and even memory. Hence the Witch was quickly 'proved' to be a widespread and extremely sinister figure with amazing powers.

The notion responsible for ready availability of proof was, like so many others, a Roman one. Although it was not permitted to torture Roman citizens, slaves involved in a judicial inquiry had to be tortured. It was believed that you could not get the truth from them otherwise. This idea lingered on somewhere in European jurisprudence, but lost its link with slavery, re-emerging as an assumption that, at least when dealing with the lower orders, getting at the truth and torture were associated. Without this idea, the Witch would probably have remained a shadowy figure of only passing interest and many fewer women (and men too – the Witch was not exclusively feminine) would have been done to death[8] over a period of three centuries lasting right up to the start of the Enlightenment.

The strangeness of the power of ideas to compel or overwhelm emerges even more clearly in relation to the Millennium than to the Dance. There are perhaps inbuilt, biological predispositions to follow-my-leader or to find release from psychological turmoil in ritualised movement. Neural hard wiring could also be blamed, perhaps with some plausibility, for the urge to slaughter-my-neighbour which was such a regularly recurring feature of millenarianism. It's harder to believe that behaviour exemplified by the Flagellants or the Anabaptists of Munster was aided in any way by the selfish gene, or any other form of biological compulsion.

The problem is that, if one appeals to biology to explain the power of each cognitive object, of which there are many thousands, one may end up having to propose implausibly large numbers of distinct biological mechanisms, as well as a range of more general ones such as 'herd instinct'. General neural mechanisms might suffice on their own for the power of a very wide range of ideas, but hardly for the highly stereotyped manner of their manifestation in widely separated times and places. Philip Ball, in his fascinating book *Critical Mass* on the application of statistical physics to sociology, economics and even bird migration, makes a similar point: '... group behaviour may not be simply a scaled-up version of individual behaviour. Rather, characteristics of the group appear which cannot be predicted from the nature of the brain's instincts alone'. In other words, regularities in the behaviour of groups of people can emerge that have no direct connection with any single person's neurology, biology, or even habitual prior intentions, just as the properties of water in bulk are unlike those of any of its component molecules individually.

It is worth looking at some more examples of cognitive objects in action to see whether they leave any room at all for the exercise of individual free will. Taken by itself, the condition described in the next chapter suggests that there is sometimes remarkably little space for individual manoeuvre once a person falls into the grip of a story that is 'doing the rounds'.

Chapter 9

WEARINESS – VICTORIAN AND MODERN

Weariness is a conscious experience, and therefore associated with special neural activity of some sort (see Chapter 2). People used to view it as a direct consequence of physical factors such as lack of sleep or lactic acid accumulation in the muscles due to exercise. But it is now looking increasingly likely that the experience is another brain 'best-guesstimate', a bit like the guesstimate of ownership of action (Chapter 5). It seems to be the brain's assessment, appearing in consciousness, of the body's physical reserves. This estimate is normally reasonably good, though there is evidence that it is usually set so as to leave quite a bit in reserve in case of emergency. We all know from our own experience that weariness takes psychological factors into account, as well as physical ones. Tiredness can vanish if we switch from doing something boring to something interesting or frightening. One might ask, 'but doesn't the fatigue disappear simply because the brain releases dopamine or noradrenaline when we get happily busy or scared?'. That would be to miss the point. There are physiological underpinnings like dopamine release associated with all feelings and experiences. Nevertheless, experience can appear to cause physiology as well as the other way round; though the latter view is the one often favoured by reductionists. In fact the two, experience and physiology, are best regarded as aspects of a *single* process. Supposing otherwise is really only a hangover from Cartesian dualism (see Chapter 2), and can lead to all sorts of futile 'chicken and egg' arguments.

The capacity to feel weary at all is no doubt hard-wired into the brain to ensure that we rest when necessary. Its sensitivity – the 'thermostat' setting as it were – and the circumstances in which it is felt are more malleable. When we have flu or the like the setting is very sensitive and doing anything makes us tired, though in fact, if our muscular strength were to be tested, it would be little different from normal. The setting can also be influenced by psychological factors and these can behave as cognitive objects, like the Noble Roman or Tarantism. The advantage of Weariness in building a theory of choice is that it has recently been scrutinised from all sorts of points of view.

In Victorian and Edwardian times, excessive weariness of uncertain origin was called neurasthenia. Nowadays it goes by a variety of names, most prominently ME (myalgic encephalomyelitis) or chronic fatigue syndrome. Neurasthenia and ME seem to be different names for identical conditions. There is a fairly recent caveat – 'Few controversies in modern medicine have raged so fiercely as that over the syndrome which has been called myalgic encephalomyelitis or sporadic neurasthenia' wrote psychologists Massimo Riccio and his colleagues in a 1992 paper on the topic. Nevertheless, most dispassionate observers nowadays view it as hardly ever due primarily to physical illness, though this can play subsidiary roles. It's the psychology that plays the major part. Perhaps there are some people whose ME is due entirely to physical illness, but none have been discovered for sure (so far). If any exist, they are likely to be rare. That's not to say that sufferers don't *feel* just as bad as someone with a virus like flu. They do – indeed they often feel worse than does a person with an equally disabling condition due to physical causes. Let's take a detailed look at the history of this Weariness object.

In the 17th century, a number of poorly defined illnesses were characterised by weakness and fatigue, which was clearly a pretty dangerous sort of thing to experience. According to John Graunt, a pioneer statistician who wrote about deaths in London (in 1662), 'lethargy' as a cause of death was about on a par with murder or starvation. (Incidentally, by far the most common cause recorded then was something called 'apoplex', which seems to have encompassed a whole range of conditions.) Young ladies were particularly prone to Fatigue, manifest in what was called 'chlorosis'; one such was described in 1697 by a physician named Pierce based in Bath, the fashionable spa town. She was 'Faint and Tyrie, upon the least stirring...'. She recovered, in due course, with the aid of the town's baths. Chlorosis overlapped to some extent with diagnoses of 'the Mother', otherwise called hysteria, but was usually thought to be a distinct condition. Incidentally, only women could be hysteric; similar symptoms in men were attributed to the spleen instead of to the womb. Then there was 'sciatica', which could prevent walking but was not necessarily painful, unlike its modern equivalent – a slipped disc – which is nearly always painful. 'Sciatica' was in fact recognised as being sometimes of psychological origin in 1961 by Robert Boyle, the great chemist and originator of Boyle's law, who was also a self-confessed hypochondriac. In short there was a concept in the 17th century that weakness and fatigue could be components of a range of diseases with rather mysterious causes.

Prior to the 18th century the epithet 'nervous' implied sinewy, strong and vigorous. From the early 1700s persons of good breeding began to suffer weakness of the nerves brought on by the pleasures and dissipations of city life, and thus became nervous in the modern sense. George Cheyne, an influential British physician, seems to have authored an idea which still has an important role, though perhaps he simply popularised a notion which was already in the air. Of 'the vapours, hysteric or hypochondriacal disorders', he wrote:

> it seldom and I think never happens or can happen, to any but those of the liveliest and quickest natural Parts, whose Faculties are the brightest and most spiritual, and whose Genius is most keen and penetrating, and particularly where there is the most delicate Sensation and Taste, both of Pleasure and Pain.

In other words only the best and most sensitive people could get knackered for no apparent reason. The treatment for such conditions at the time was a good dose of plain country living, fresh air and regular exercise. Hence the concept did not enter into a particularly important alliance until the following century, as is recounted below, when the situation was more favourable.

At the start of the 19th century Thomas Trotter, a naval surgeon and then a physician in Newcastle upon Tyne, far from fashionable London, wrote:

> Sydenham at the conclusion of the seventeenth century, computed fevers to constitute two-thirds of the diseases of mankind. But, at the beginning of the nineteenth century, we do not hesitate to affirm, that nervous disorders have now taken the place of fevers, and may be justly reckoned two-thirds of the whole, with which civilized society is afflicted.

Mind you, he was not necessarily the most reliable of guides, since he had in his naval days resisted the use of lime juice for the prevention of scurvy. Nevertheless, he was expressing what must have been a fairly widespread view, one consequence of which was that nervousness rapidly lost its appeal to exclusive society and to the 'sensitive'. Many of them became quite hostile to any suggestion that their own afflictions might be of this sort, just as, prior to beguilement by Cheyne's claim, people thoroughly resented any implication that they might be hysterical or hypochondriac. Around the same time the bleedings, purgings, vomitings, cuppings and shockings of the 18th century gave way to less brutal prescriptions in almost every branch of medicine; among these was the notion that prolonged rest could often effect a cure.

Thus, midway through the 19th century, a group of concepts that were soon to combine existed in a sort of limbo. They were:

(a) weakness and fatigue are symptoms of some poorly understood illnesses
(b) there are illnesses to which only particularly talented and sensitive people are prone
(c) illnesses of type (b) are not 'psychiatric'
(d) rest is a good cure for many conditions
(e) much disease is due to environmental causes.

This last derived from numerous writings attributing illness to fashionable life, a 'civilised' way of living, appalling slum conditions associated with the Industrial Revolution, and so on. Extra impetus was soon added to the idea by the mid- to late 19th century discoveries about bacteria in the environment, and the consequent germ theory of disease.

The American neurologist George Beard, writing first in 1869 and then at greater length in 1880, popularised a disease called 'neurasthenia'. Allegedly he suffered from it himself. It was a composite of the five concepts listed above. Profound fatigability of mind and body were the main symptoms; it was more common among educated and professional classes; it was a physical, not a mental, state; the treatment was rest; it was largely due to environmental factors. Despite meeting scepticism at first, the disease quickly became widespread in parts of America and Europe, though it was never quite as popular in England, possibly because physicians there on the whole remained sceptical longer.

A similar disease, which achieved a more limited popularity at around the same time as neurasthenia, was 'effort syndrome', described by another American, Jacob Da Costa. This was a little different in that it affected the physically fit rather than an intellectual or social elite. But these patients, too, should on no account exert themselves, doctors said. It was due to a specific, though undetectable, physical lesion of the heart; the fifth concept concerning environmental causes was, in other words, absent from the idea of effort syndrome. It resulted in fit young men, especially those liable to be conscripted for the

Spanish–American war, being overcome with exhaustion on minimal exertion. It was not diagnosed solely by army medical officers selecting recruits, but also cropped up in civilian life.

The component concepts of effort syndrome are specific variants of four of those contributing to neurasthenia, rather as some of the ideas belonging to Tarantism were more specific than those of St Vitus' dance. This should have made effort syndrome more persistent and less 'epidemic' than neurasthenia, and probably would have done so but for the alleged subtle heart lesion. Any such lesion eluded detection despite the introduction of increasingly sophisticated methods for uncovering heart disease. As one of its core concepts dissolved, the syndrome crumbled despite a brief revival as recruitment for the First World War gathered pace.

Neurasthenia suffered a similar fate; the core concept to crumble in its case was also the notion of its non-mental origin. Several sources tried to explain how disabling neurasthenia could be compatible with an appearance of perfect physical health. Their efforts resembled those of Ptolemaic astronomers adding epicycles to make their system work. The neurasthenia theoreticians soon took a step too many in adding a concept of 'functional' physical disability. This had such strong resonance with the recently developed notion of functional neurosis as to prove fatal to the diagnosis. The core concept of physical basis crumbled and the residue of neurasthenia became identified with neurosis and hysteria, and lost all general popularity on coming to share in the disapproval of neurosis associated with the First World War. In 1939, at the start of the Second World War, there were still 100,000 English First World War pensioners whose disability was recorded as neurasthenia. The number diagnosed with effort syndrome was less than 20,000. Neurasthenia was not diagnosed during the second war.

The idea of neurasthenia had been in poor health for some time before the *coup de grace*. People had begun to point out that it was at least as prevalent among lower social orders as among elites;

they had also voiced disillusionment with the value of rest cures. After the concept finally succumbed, scattered new cases of fatigue syndrome continued to occur, but they were no longer given a special name, even though pensioners with chronic illness were not deprived of their first war diagnoses. One important aid to the spread of the idea – it having a special name – was removed and, perhaps as a result according to Simon Wessely, a contemporary British psychiatrist who has made a special study of the condition, '... it is plausible that the illnesses represented by neurasthenia... were actually less prevalent during this period'. In the 1950s it returned, in a new guise, with a bang.

The object returns

When Fatigue came back, it did so in epidemic form. Although people first clearly realised only around 1955 that something very peculiar was going on, it later became obvious that epidemics had been occurring for about two decades before then.

The disease attracted many names – e.g. abortive poliomyelitis, Royal Free disease, chronic fatigue syndrome, post-viral fatigue syndrome and (benign) myalgic encephalomyelitis ('benign' was dropped when the realisation dawned that it wasn't). Many people have now settled for the last of these, shortened to ME. Doctors on the whole prefer the term 'post-viral fatigue syndrome', or sometimes 'chronic fatigue syndrome' nowadays, but this was not the case earlier on. These concepts overlap with ME, but are even more poorly defined. The 'post-viral' epithet is usually reserved for conditions at the milder and less chronic end of the scale.

The information in Table 1 is culled from a monograph by a doctor, one A. M. Ramsay who was involved in treating many of the 1955 cases in London. He became a firm supporter of the hypothesis that they were due to a virus – i.e. that they had an environmental cause, the fifth concept of neurasthenia. Outbreaks popped up all over the developed world. England appears

Table 1 Outbreaks of probable ME

Date	Place	People affected
1934	Los Angeles	198 doctors and nurses
Jul. 1937	Switzerland	130 soldiers; 28 nurses
Sep. 1939	Switzerland	73 soldiers
1948–49	Iceland	High school students
1949–51	Adelaide, Australia	700 people
1950	New York State	'Many' people
1952	Middlesex Hospital, London	14 nurses
Jul. 1953	Washington DC	25 nurses plus 25 others
1953	Coventry, UK	'Some' nurses
Feb. 1955	Cumbria and Durham, UK	Primary school children
Feb. 1955	Durban, South Africa	98 nurses
Jul. 1955	Royal Free Hospital, London	292 staff, 12 patients
1958	Athens, Greece	27 trainee nurses
1959	Newcastle, UK	Trainee teachers
1961	New York State	15 nuns
1964–66	Finchley, London	370 'cases'
1970–71	Great Ormond Street Hospital, London	145 staff, mostly nurses

to have been especially favoured, but this may have been simply because the National Health Service at the time was better at recording unusual occurrences than services elsewhere. Men and women seem to have suffered more or less equally, except when some special occupational group such as soldiers or nurses was exclusively affected, but the condition rarely attacked the middle-aged or elderly. Professionals were especially susceptible, nurses and doctors being the most vulnerable. Interestingly enough, around 10% of George Beard's original group of neurasthenics were also medical doctors – himself among them.

The outbreak at the Royal Free Hospital in London attracted widespread publicity; it affected 292 hospital personnel, but only 12 patients. It is a tribute to the power of ideas that anyone aware of this should remain firmly convinced that the epidemic was due solely to a viral cause. Ramsay does not attempt a direct explanation of how viral infection could have accounted on its own for the peculiar vulnerability of staff. He merely hints that the sporadic cases, which were known to be occurring in London at about the same time, infected medical personnel, who proved contagious to other staff while unaccountably failing to infect their patients. Staff dining rooms or naughty goings-on in nurse's hostels usually get the blame in such circumstances, but they were not incriminated in this case.

The symptoms were of a relatively minor flu-like illness, except that many cases did not have fever, while muscular fatigue was usually disproportionately severe. There were often symptoms of nervous system involvement, including patchy numbness and weakness, emotional lability and mental exhaustion. Nobody died and most cases improved with time, but a large proportion of those affected in most outbreaks remained unwell for months or years. The main residual symptom was usually abnormal tiredness following any sort of exertion, which continued to plague as many as 75% of people included in some outbreaks and was sometimes of crippling severity. Although epidemics appear to have ceased after 1971, sporadic cases continue to occur and fatigue remains their most distressing difficulty. One cannot put a figure on the exact number, since the condition has so many alternative names, while fashions in naming it change from year to year. All the same, ME and related conditions may be as prevalent now as neurasthenia was in its late 19th century heyday.

The ME epidemics were probably due to a climate of realistic apprehension engendered by polio epidemics, as Simon Wessely has proposed. Before polio vaccination became available at the end of the 1950s there were frequent outbreaks of this terrible disease, which often killed or crippled and whose main symptom was muscular paralysis. Most people in the Western world knew someone who had died of it or been permanently disabled. Although soon shown to have nothing to do with polio virus, some early attacks of ME were diagnosed as 'abortive polio' and, in the 1949 Australian outbreak, an ME epidemic followed on directly from a polio one. This connection may also account for some of the special vulnerability of hospital staff, who bore the emotional brunt of looking after polio victims.

The Great Ormond Street Hospital outbreak of 1970 happened a bit too late for polio to have been a significant factor; hospital staff must have been vulnerable for other reasons. The concept that elites are especially vulnerable is a component of the Weariness object; medical and nursing staff in general are special people in this sense, while Great Ormond Street in particular is a world-leading institution. These people may have succumbed because they fitted the 'profile' of Fatigue particularly well.

Myalgic encephalomyelitis

By 1993 the eminent psychiatrist Bob Kendell, a professor in Edinburgh at the time, pointed out that doctors were little nearer to understanding the causes of ME than Beard, the neurologist who first described 'neurasthenia', had been a hundred years ago. That said, most doctors of the time would have agreed with the following:

- ME causes much distress and severe incapacity to many of those afflicted.
- It is not due to malingering.
- It is often chronic.
- Although most sufferers improve with time, fewer than 10% of some groups studied recover completely within three years.
- A large proportion of people experience fatigue that they regard as excessive but very few attribute this to ME.
- The onset of ME is sometimes, but not always, precipitated by a virus infection.
- Some cases, but not all, show immune system abnormalities consistent with chronic viral infection.

- Early evidence that Epstein–Barr or Coxsackie viruses are especially implicated has not been reliably confirmed.
- Some cases have symptoms of depression; a small proportion improve with antidepressant medication.
- Objective evidence of physical or mental weakness is not always to be found; when it is, the subjective symptoms patients describe always appear disproportionately severe.

Sufferers from ME themselves were naturally less dispassionate: 'There are at least 150,000 people in Britain today who are suffering from a strange illness which nobody understands. This illness can have a devastating effect on one's life'. So wrote Anne MacIntyre in 1989 in the opening of her popular book. Like Beard, she knew only too much about it from personal experience. She listed ideas about causes: 'Virologists find ME is usually caused by a virus. Immunologists find ME is a disordered immune system. Clinical Ecologists say ME is a condition of multiple food and chemical allergies. Psychiatrists decided (many years ago) that ME is a psychiatric illness – i.e. we are mad. Neurologists think ME is a disorder of the nervous system'.

The professional groups that MacIntyre mentioned all tended to think that they did not fully understand ME, though each may have had something to contribute. Her tone of approval for environmental explanations, but indignation over a psychiatric one, found a powerful voice in the ME Association. This was a UK-based organisation, recruited almost entirely from sufferers from ME or their close relatives, which promoted the idea that ME is a specific disease entity with environmental causes (equivalent organisations exist in many other countries, where they often call themselves ME Societies). It provided 'fact sheets' for patients and doctors telling them what symptoms patients could expect and what treatment they should have. The main treatment, naturally, was rest.

The Association was a remarkably effective pressure group; it even got an 'ME Act' made statutory in England. This required that an annual report be made to Parliament on the causes, effects and treatment of ME. Moreover, it managed to get ME included under 'diseases of the nervous system' in the International Classification of Diseases, thus firmly implying organic causation. And, according to a 1994 editorial in the *British Medical Journal*, the Association kept up a steady campaign of disparagement against anyone suggesting that the condition might have psychological facets.

It was almost as if the idea of Fatigue had taken on a life of its own, having learned from the failures of its previous incarnations as neurasthenia and effort syndrome. Both the previous conditions disappeared only when belief in their physical basis failed. The ME Association concentrated many of its efforts on preserving this particular concept, and showed much more flexibility when it came to discussion of what environmental causes might be important, or what treatment measures in addition to rest were effective. Neurasthenia crumbled only after it had been weakened by damage to two core concepts: beliefs in the greater vulnerability of 'special' people and in the value of rest as a cure. The ME Association, in the present social climate, was hardly able to defend outright a view that only 'the best' people were particularly likely to become ill. This was implicit, however, in some of the literature for the general public that they approved, and indeed in one of the popular names given to ME in the 1980s – 'Yuppie flu'. As it happens, the condition is probably now blind to social status. By itself, the fact of spread through the whole social gamut does not necessarily undermine the idea, as we are all special people to ourselves, but has probably contributed to weakening it.

There is now a fairly general medical consensus that introducing graded exercise is the best treatment for ME. When the evidence was beginning to mount that rest was not advisable, and may indeed have contributed to some of the symptoms and to any abnormal physical findings, the ME Association adopted a 'yes, but' strategy: 'Yes, but rest is essential in the early stages; yes, but too much exercise can cause severe relapse'.

The whole situation seemed very confusing ten or twenty years ago. Now it is possible to draw a sketch of what has probably been going on. The cognitive object gains a foothold in people whose vitality is temporarily lowered for everyday reasons, or because of emotional stress, minor infections, an attack of depression or other adversity. It is impossible to prove that ME is never due to organic disease unknown to medical science, so maybe there are a very few cases whose condition was or is primarily due to illness of this sort. The idea found circumstances congenial to its epidemic spread among relatively sophisticated people who were aware of the relevant concepts, during all the anxiety created by polio epidemics, polio being an illness with superficial similarities to ME. Rare sporadic cases had always occurred, especially in high achievers who had difficulty in tolerating the everyday weaknesses to which the flesh is heir. When polio scares subsided, cases of ME might have been expected to remain rare or to fade away altogether. That this did not happen seems to have been largely due to the tireless efforts of some sufferers, joining the ME Association and making use of their capacity for high achievement, to propagate the relevant concepts and ensure their availability to everyone.

In 2004 the *Journal of the Royal Society of Medicine* published a long-term survey of about 2.5 million UK residents, comparing their experiences in 1990 and in 2001. In both years around 1.5% of them told their family doctors they had developed excessive fatigue in the preceding year. The proportion of these people given a formal diagnosis of any type of fatigue syndrome almost halved over the decade, from about 58% of those reporting abnormal fatigue to about 33%. Most of this decline was in diagnoses of 'post-viral fatigue syndrome' (the poorly defined condition, not clearly distinguishable from ME, which is generally deemed to be milder and less chronic). The overall number of specific diagnoses of ME actually increased slightly, suggesting that the idea is still very much alive and active in some medical and other circles.

Why should the condition be so chronic in many people? The 'rest' concept plays a crucial part here. Anybody who has been kept in bed for a week by illness or surgery knows that one quickly becomes feeble through lack of exercise. Some ME cases hardly moved for months on end, so naturally they felt extremely tired and started to ache when they tried to do anything. On top of this they were often told (by books or pamphlets on ME) that they would suffer terribly if they did too much. Experience quickly taught them the truth of this. Much the same used to happen to heart attack cases in the days when they were threatened with death (by their medical advisers) if they did not keep to their beds for three months. The policy then changed to getting heart patients up and about as soon as possible, usually within a few days of their attack. Fewer die now, and far fewer become chronic invalids. The same policy applied to ME could prove equally helpful.

Y

This sorry tale illustrates how cognitive objects can behave. A group of seemingly unrelated concepts can suddenly coalesce into an idea that, to all appearances, takes on a life of its own. People are swept up into it with little or no choice in the matter. From the earliest days of neurasthenia to the present, Fatigue sufferers have described how they have felt overwhelmed by something outside of themselves. Moreover, in this particular case, those affected are often people especially likely in normal circumstances to feel in control of their own lives and fates. In a sense Fatigue is like an ongoing story that has taken over substantial parts of the life stories of individuals, quite against their wishes.

Next let's take a look at how concepts can behave when they are invited in, so to speak – when people deliberately choose to adopt them. The concepts chosen have the further characteristic that any connection they may have with brain 'hard wiring' is likely to be a lot more remote than is the case with Weariness or, to a lesser extent, the Dance.

Chapter 10
GOOD IDEAS

Saints are probably the most obvious example of people who actively strive to incorporate a cognitive object within themselves, so it is worth taking a look at how some of them got on. The actual idea of the Saint is quite simple compared to many, so we'll also examine a couple of more complex ideas – a personal one, the Doctor, and an impersonal one, the Hospital – to try to get an impression of how these have affected individuals. Please remember that I'm writing about the make-up and behaviour of these objects, not about their truth or other value. Those matters are left to each reader's judgement.

Various saints

Christian saints are, in principle, with exceptions to prove the rule, people who have conspicuously lived up to Jesus's two commandments: love God as intensely as possible, and love your neighbour at least as much as yourself. The 'at least' is there because some saints have had little overt love for themselves, and would have been open to criminal charges if they had shown equal regard for their neighbours. Exceptions proving the rule have been mainly people to whom very intense love for God could be attributed despite absence of evidence for charity or other virtue in their worldly dealings. St Wilfrid, whose main claim to fame was winning an argument with the Celtic church at the Synod of Whitby in 664 about the date of Easter, was an exception of this sort. He was an energetic but quite unpleasant character, prone to quarrel with all and sundry. In fact, a good many of the earlier saints seem to have been rather similar to

Wilfrid in personality as well as behaviour. In addition to evidence of love for God and (sometimes) neighbour, there must also be proof of divine endorsement for sanctity in the form of miracles associated with the saint before or after death. Whatever one's view of the status of miracles, it is at least clear that saints must have made a deep impression on people in order to have such marvels attributed to them.

St Thérèse of Lisieux[1] was a particularly simple saint. Born in 1873 to bourgeois parents, she died in her convent aged 24. She was the youngest of four girls, affectionate and intelligent. Both parents had themselves wanted to enter religious orders in youth but had been turned down. They remained religious to a fault, attending mass every day throughout their lives at 5.30 a.m. Before she was three, Thérèse reportedly said that she wished her mother could die so as to be in heaven. Mother became ill soon after and did die around two years later. Thérèse had to go to a foster home for a while before the death, but was subsequently brought up by her father, to whom she became very attached, and her older sisters. Two of the sisters then entered the local convent, a strictly enclosed Carmelite order, and Thérèse became depressed for some four years. She recovered after a statue of the Virgin Mary in the house had come alive and smiled at her. Many people under stress have similar experiences. Her hallucination was not an indication of serious mental derangement.

She had long been aware of a vocation to become a nun and of a capacity for sanctity of a sort. It is a measure of her strength of character that she got the authorities to bend the rules and let her in early when aged only 15. Her father died soon after she completed her novitiate and the remaining sister, who had stayed home to look after him, joined the rest of the family in the same convent. Thérèse spent nine self-denying years enclosed there before dying slowly and painfully, probably of tuberculosis. In the meantime she wrote her famous *Diary of my Soul*, an exercise originally suggested by one of her sisters in the hope that it might be therapeutic. The diary's saccharine spirituality ensured its widespread popularity, and soon led to Thérèse's canonisation.

It's a pitiful story. One's heart aches for this poor child who suffered so many losses and never, it seems, quite grew up. Her undoubted courage and other virtues always retained a quality of the good little girl anxious to please the grown-ups, who in her case were represented by her concept of Jesus. What did it look like from her point of view? There were certainly rationalisations to compensate. Of her father's own numerous losses, she said: 'Our father must be greatly loved by God, since he has so much to suffer. What a delight to share in his [God's] humiliation'.

Of herself she wrote:

> I am happy, yes, truly happy, in having no consolations. I should feel ashamed if my love [for Jesus] resembled that of earthly fiancées who look for presents from the hand of their betrothed, or eagerly watch his face for the loving smile that delights them... (translation by M. Hollings)

Yet these defences were far from impregnable. When dying, for instance, she wrote, 'Press on, press on, looking forward to death! But it won't give you what you hope – only deeper darkness still, the darkness of extinction' (M. Hollings).

Had she not been so wholly occupied by the idea of the Saint she might have sunk into despair or delinquency, or she might have matured. She might even have avoided adding to her father's losses – clearly something that she would have chosen to avoid if she could.

Sorrow and pity are about the last emotions inspired by St Francis of Assisi's story. He was originally a romantic dreamer of the sort only too liable to stagger from disaster to disaster; fortunately a rescuer was always at hand. Before he was 25 his hard-headed father usually filled this role; subsequently God did. Probably the most painful time in his life was the transfer from the protection of one to the other. His luck did not fail him even then as he discovered a temporary passion for repairing derelict churches with his bare hands, which served to keep his mind off things until God was fully in the saddle.

His message consisted of taking a few ideas from the New Testament literally – 'Sell what thou hast, and give it to the poor, and thou shalt have treasure in heaven', 'Take no thought for the morrow' – and ignoring the rest of the scriptures. His method, applied rigorously, resulted in a life of social parasitism for himself and his followers, who were soon numerous. A central message of his early preaching was, of necessity, that anyone who gave him a few morsels of food now was buying heavenly treasure worth a thousand silver marks; a piece of sophistry that only a future saint could use without attracting accusations of dishonesty. His followers, not being saints, were left in a difficult position which they eventually resolved in the only way possible: although individual Franciscans remain frugal, there order is now ell endowed.

Francis's treatment of women and his family is interesting. Basically, he would have no truck with either group once God had taken over. Fornication was, in his view, one of the most serious offences possible to a Franciscan, meriting instant dismissal without possibility of repentance. He lived up to his own precepts in this, as in all other areas of his life once he had really got into his stride. He avoided his friend St Clare, who loved him, until the very end of his life, when he allowed her to nurse him through his last few weeks. He was quite right to be so cautious, of course. In someone as warm-hearted and impulsive as he, sexual love, children and family life might soon have displaced the Saint from his mind. All the same, his total involvement with the idea produced altogether remarkable results. He changed from an affable drifter into a virtuous hermit, who spent most of his final years on a private mountain that had been donated by an admirer. There is no reason to doubt that he developed stigmata (the wounds of crucifixion) and may have been the first to do so. And he has been remembered, for 800 years now, as one of the greatest and best loved of the saints.

A less colourful sainted Francis, François de Sales, lived about four hundred years later (1567–1622). He was a sprig of the minor nobility, a sort of colonel in the army of the late counter-reformation; his first mission was to re-convert the Protestants of his native Savoy. At an early age he was made bishop of Geneva – not the easiest of appointments as the town had long been Calvinist. He beavered away in parts of his diocese that were still Catholic, became a popular preacher and produced high-minded writings which appealed to his contemporaries, though they have not lasted well. Probably his best book was *Treatise on the Love of God*. The picture is of a kindly but obsessional man of smug piety; the early 17th century equivalent of today's well-heeled do-gooders.

By his late thirties he had long been a willing provider of spiritual direction for ladies in whom he discerned special merit. It seems to have been a titillating hobby disguised under thick coatings of pastoral propriety. Then he fell deeply in love with one Jeanne de Chantal. Luckily, just like Francis of Assisi's Clare, she too was sainted in due course. Propriety was maintained, with difficulty. Though slightly younger than de Sales, Jeanne was already a widow when she met him, with four children and a very difficult father-in-law. Worse, she lived a long way from François' home. Examining the state of her soul via the post proved insufficient for both of them. Quite soon she was looking after his youngest sister, despite the fact that his mother was still alive and well. Then one of his younger brothers found himself married to one of her daughters. Tragically, the sister died while in Jeanne's care and both of the surrogate newly-weds also died young.

Surrogates had proved inadequate, however, even before the second tragedy. They decided to follow Clare's example, or that of Heloise after Abelard's castration. A nunnery was the thing, but it had to be a very special nunnery with not too strict a rule, not enclosed and near François' home. As he had spent a good deal of energy in the recent past reforming his local convents, and making sure that they stuck to their rules, it was a fix. God provided the answer: it turned out to be His Will that Jeanne should head a new order with a rule, to be concocted by François, suitable for the delicately nurtured. God also inspired a coat of arms for the new foundation. It contained a bleeding heart, pierced by two (presumably Cupid's) arrows. Quite soon Jeanne had left her remaining children and troublesome father-in-law, and was installed within walking distance of François in a brand new nunnery. She was in charge; most of the early, subordinate recruits were ladies who had previously benefited from François' spiritual direction.

The new order soon proved a great success, perhaps because it had been designed with comfort in mind. Jeanne was ever more occupied with her management role as it expanded, but François liked to sit in the original foundation surrounded by ex-directees for whom he provided uplifting thoughts. In the words of a 20th century biographer: 'In all the annals of the saints there is perhaps no more delightful and unforgettable a picture than that of François de Sales sitting of a summer's evening in the orchard of the Visitation by the lakeside, surrounded by the little company of eager sisters as they listened to him chatting to them of all that in the life of a religious community or indeed in any human life goes to make up a life of the love of God in response to God's love of us'. Of such homilies were François' books made; given extra impetus, in all likelihood, by sexual frustration as he probably stuck to the letter of his vows.

There's a certain contrast evident in the quality of sanctity shown by Francis of Assisi and the other two. Maybe this could have been due to the radical change wrought by the invention of printing in the way concepts could spread. Thérèse and François can be regarded as saints of the word; their reputations depend mainly on their own writings, though of course the lives that they led had to show at least a surface consistency with what they wrote. Francis, on the other hand, was of necessity a saint of the deed; his life had directly to create a great impression on others before they would bother to go through the laborious process of making manuscripts about it. If the change in how ideas could spread was sufficient to explain the apparent change in the nature of saintliness, then other later-born saints should also be more like François than Francis. But this turns out not to be always the case.

A saint, born after the invention of printing, who was noted to be 'always something more and something greater than what he said or wrote,' is Ignatius Loyola from the Basque country of Spain. Belonging to the squirearchy, Ignatius was expected to be a combination of knight errant and self-employed officer, possessing both civil and military management skills useful to royalty or other patrons. In his youth the romances and songs of the troubadours, long out of favour in France and England, were still popular in Spain and moulded the imaginations of people of his class. Ignatius (then called Inigo de Loyola) threw himself into the expected role with zeal. He was reputedly a quarrelsome fellow, brave, ambitious, quick to draw sword in defence of his honour, prone to wenching, in love with an unattainable lady from the royal household – and not very literate.

Aged about thirty, he found himself trying to inspire a reluctant garrison to defend the city of Pamplona against a French army. It was a quixotic act, as the citizens surrendered immediately, having had past experience of the ill consequences of provoking a siege, and the situation was hopeless without their cooperation. All the same, Ignatius led a brief resistance and a cannonball 'passed between his legs' wounding him seriously. He was allowed home to the family castle after first aid from a French surgeon, and needed a series of operations on his right leg which kept him bed-bound for nearly a year.

In the castle were (only?) two books; a romanticised life of Christ and an equally high-flown account of a selection of saints. Ignatius spent his year ploughing through, and meditating on, these two works. At the end of it he was a half-changed man. He left the castle, clothed as a pilgrim, intending to go to Jerusalem. His half unregenerate status showed itself when he fell in with a Muslim travelling in the same direction. The Muslim allowed that the Virgin Mary was a virgin immediately before Jesus' birth but would not agree that the same applied after delivery. He rode on fast when Ignatius' temper began to fray. Ignatius was in two minds about whether to pursue and knife him 'in defence of Our Lady's honour'. In the end Ignatius's mule made the choice: it decided, when allowed its head, to take a different route from the Muslim.

The rest of the change in him occurred in the course of the following year, during a prolonged stopover near Barcelona spent mainly in prayer and penance. Here he wrote his famous *Spiritual Exercises*, a short guidebook based on his own experiences of how to go about achieving knowledge of, and allowing reform by, God. At the end of this second year he knew that he was to form a company, later known as the Jesuits, devoted to promoting the honour of God. First, however, he had to fulfil his vow to visit Jerusalem and then remedy the many gaps in his education by going to university.

The pilgrimage did not take very long, but it was another ten years before he completed his education. He did not have much luck with Spanish universities as the Inquisition kept putting him in prison and interfering with his extra-curricular preaching activities. In the end he went to Paris and got a master's degree there. There also, he gained companions who went through the spiritual exercises and came to regard him as their leader.

In the last years of his life he continued to get 'illuminations' (cognitions coming directly, he thought, from God) and visions of the Trinity, but spent most of his time in an office in Rome writing endless letters of guidance, encouragement and administration for the ever-expanding Jesuit order which had spread, before he died, from Japan to the Americas. He had tried to escape this fate, but the companions had insisted on electing him 'General'. Perhaps early officer training was too ingrained to let him follow Francis's example and retreat to a hermitage on his own.

He seems to have been more competent than either Francis in dealing with women. Early experience may have helped him in this respect, even if it was a hindrance when it came to escaping duty. He was handsome and charismatic, and attracted female devotees. Several of those that he acquired during his Barcelona period remained on good terms with him all his life. He dealt with them affectionately, apparently without the favouritism shown by François or the fear that Francis must have experienced. He also put a lot of effort into helping prostitutes without, it seems, the motivational ambiguity that Mr Gladstone showed in the same endeavour[2]. However, he refused to let women join the Jesuits, though one of the Barcelona ladies got round this by appealing directly to the Pope. She later accused Ignatius of stealing her money, but an independent inquiry showed that in fact she was in debt to him and the Pope was persuaded after all to support Ignatius's line. It's a measure of his sanctity or her genuine devotion (or both) that she allegedly 'returned to Barcelona without bitterness'.

An anatomy of sainthood

Clearly saints are preoccupied with ideas about God and love, often to the exclusion of everyday concerns. Concepts of God's perfection tend to highlight their own relative unworthiness, leading some into excessive penance. Ignatius is said to have permanently undermined his health through self-mortification of various sorts while at Barcelona, but later tried to discourage others from following his example in this respect. Few have succeeded in reconciling an ideal charity with more personal, particularly sexual, love. Thérèse tried to reach the ideal by etherealising what she knew of physical love. Only her ignorance of the latter allowed this. Francis's approach was more typical; he simply turned his back on any possibility of sexual entanglement. François seems to have got caught, just like many a modern psychotherapist falling in love with a patient. Ignatius cultivated his talent for friendship and tried, mostly successfully, to generalise it. Even he, however, could not cope with women as colleagues. The fact is that love of God, if taken to the extremes that saints wish, is not compatible with any exclusive, biologically based love of an individual, nor even with consistently adequate self-care. The necessary concepts and behaviours are to some extent mutually exclusive.

Saints' relatives tend to get a raw deal. Francis cut himself off from his parents, though his father had been supportive, if interfering, while his mother was apparently devoted to him. He added insult to injury by publicly stripping himself naked when, during a quarrel, his father asked for the return of money and clothes that Francis had stolen from the household. Thérèse seems to have knowingly added to her father's many crosses, and it can't have been that easy for her older sisters either, having to share a convent with an incipient saint. Ignatius' proud family did not take kindly to his conspicuous poverty. Even François, though always kind to his mother, colluded with his fellow saint, Jeanne, to discover God's wish that she should abandon her father-in-law and teenage children. Especially in his case, there was an obvious conflict between personal wishes and the ideas embodied in the Saint, which led to choices that look bizarre at best from a modern perspective.

The early stages of sainthood are understandable in terms of trying to live up to a set of concepts, or a sort of story. The source of the ideas was particularly obvious in Ignatius's case. He was in a vulnerable condition, due to personality and circumstances, when he suffered intensive exposure to notions about sanctity stored in two books. Thérèse acquired the ideas from her parents. The other two proved susceptible to an aspect of the general culture of their times. Once the Saint got a foothold in their minds, however, it appeared to behave rather like Fatigue in that it took on a life of its own, albeit with its host's conscious collusion. For instance, the importance given to the subsidiary concept of chastity helped to prevent the Saint being displaced by mundane concerns such as marriage or fatherhood, just as the notion that 'rest is best' helped Fatigue to keep going. At least in François' case, chastity was clearly something that did not accord with his own deepest wishes. As in the case of Fatigue, the Saint, too, found strategies for ensuring that multiple copies of its core concepts were made and placed in books or in organised groups of people where they would stand a good chance of influencing future generations.

At a later stage, though, something other than mere story-telling appears. The individual saint becomes an integrated embodiment of the story, rather than just someone who endorses its component concepts. Thérèse and François may not have fully emerged from the condition of entertaining separate concepts as opposed to *being* the Saint, but the other two did. The example of Ignatius suggests that the difference was not due to the invention of printing[3], as he was closer to Francis in quality of sanctity but was (ultimately) the most book-learned of the four. If one believes Marshal McCluhan's dictum that the medium is the message, it appears that the 'medium' for the Saint story is not the literature in which it is told but, rather, the people in whom it is embodied.

Healers and doctors

St Teresa of Avila put the matter succinctly. 'Healing people', she said, 'is my recreation'. The miracles of saints often centre around mysterious cures, but this is a sideline; the core concepts of sanctity deal with very different concerns. The King, too, encroached on the territory of the Healer, as in the king's touch, which was supposed to cure scrofula (tuberculosis of lymph glands under the skin). George I evaded this regal duty; when asked to heal 'the scrofulous son of an English gentleman', he advised the man to go off to France and try the Stuart pretender instead. The man did so, the son was duly cured and the Stuart cause gained two supporters. Despite the existence of trespassers like these there are others whose main business is healing.

The pure Healer is a rare and elusive figure, sometimes represented in lore and literature as an innocent farm hand or middle-aged housewife who discovers quite accidentally that they have a gift for curing people, often through touch. Because this phenomenon is elusive, and society needs a regular supply of healers, a very different type has developed to supply the need –

the Doctor. Many doctors do have a bit of the pure Healer in them but, especially nowadays, this is hardly admissible as other component concepts are thought so much more important and respectable. The Doctor is in fact the most complex cognitive object that we have yet encountered as it is made up of many threads, most of them quite antique yet still essential.

The Hippocratic Oath and its updated version (the Geneva Declaration) relate only to doctors' behaviour, but both contain ten separate injunctions, each invoking a number of complicated ideas. The first item of the Geneva declaration, for instance, is, 'I solemnly pledge to consecrate my life to the service of humanity'. The 400-page manual which would be needed to pin down precise notions behind this statement, and elucidate their practical application, is not supplied. 'The enduring factors in medicine have always been compassion, pity, care and love'; so wrote Dr Philip Rhodes, Dean of Southampton Medical School in the 1980s. Recognising that these are not always prominent attributes of doctors in general, he went on: 'These make *nursing* [my italics] the really fundamental aspect of health care…'. All the same, compassion and the others do come into doctoring, though not necessarily as core concepts. A charming 1993 book from Yale University, *Empathy and the Practice of Medicine*, argues that they should have an essential place. That the argument needs to be made, however, indicates that they are not central. Even healing may only be clinging to the core by the skin of its teeth: 'Our technological triumphs have made many doctors feel so much like mere conduits of power that they no longer think of themselves as healing agents', says the empathy book. Actually the feeling described here would have been entirely familiar to any priestly healer in the last ten thousand years or more, except that he would have felt himself a conduit for the power of the God rather than that of a giant pharmaceutical firm.

Which, then, are the core concepts of the Doctor? Apollo, Aesculapius and centaurs were mythical physicians of antiquity, as later and in actuality were Anglo-Saxon 'leeches' for instance, but their attributes are not clear-cut except that they made people better. The original doctors were shamans or witch-doctors. John Camp, who was not a doctor himself, rather flatteringly described them in 1978 as 'usually the most intelligent member of the community – and the most feared'[4]. This element of intelligence and, not necessarily benign, learning has been a persistent thread throughout the history of the Doctor. Imhotep, the first recorded Egyptian physician, who lived around 3000 BC, was also a magician, priest and architect. Greek physicians, following the example of Hippocrates who could and probably did debate on equal terms with Socrates, tended to be among the most learned people in their society. Even Anglo-Saxon leeches had to be literate and produced medical textbooks such as the *Leechbook of Bald*, some time in the 9th or 10th century. Romans had in the meantime introduced a slight hiccup into this concept of the physician, whom they originally held in low esteem. To them, the physician was a rather unimpressive artisan. Indeed, Cato the Elder would have nothing to do with doctors, preferring to rely on raw cabbage instead. His hesitations were soon to be swept away by the tidal wave of Greek influence.

The revival of Greek medicine in Europe came via Islamic translations, by which time it had got attached to astrological ideas; in the early Renaissance any high-class physician had also to be an expert astrologer. Many added astronomy, alchemy and a little chemistry to their basic qualifications. By the 12th century the first European medical school to open since Roman times (at Salerno in the 9th century) required students to study logic for a year before going on to five years of medical studies. Incidentally, the undergraduate medical course has remained at around five years ever since – some two years longer than most undergraduate studies. The Doctor idea is full of instances of simple concepts like this that have stuck to it over centuries for no obvious reason. The arduous, and indeed esoteric, learning concept is still there, of course. Doctors in training must acquire more information about a wider range of subjects than any other profession – an odd requirement, it might be thought, given that the rate of scientific progress is now such that a significant proportion of the so-called facts they so laboriously acquire will be shown up as either irrelevant or incorrect before they are halfway through their careers. I nearly failed an oral examination at medical school because I did not know the 'fact' that cerebral blood flow remains unchanging in nearly all circumstances. Nowadays, of course, brain imaging techniques of various sorts depend entirely upon the alterations of blood flow that are constantly occurring.

Another group of concepts at the core of the Doctor idea were caricatured by Chaucer:

He did not read the Bible very much.
In blood-red garments slashed with bluish grey
And lined with taffeta, he rode his way;
Yet he was rather close as to expenses
And kept the gold he won in pestilences.
Gold stimulates the heart, or so we're told.
He therefore had a special love of gold.

The basic problem was, and is, that money is needed to fuel continuance of the Doctor, but some concepts (compassion etc.) commonly associated with it are not readily compatible with the idea of financial gain. Therefore the idea that 'the physician is worthy of his fee' became a core concept of the Doctor at a very early stage. Nevertheless it continued to cause unease. A book, *The Precepts*, written soon after the time of Hippocrates, urges doctors 'not to be too grasping, but to consider your patient's means. Sometimes give your services for nothing...'. The Dark Age historian, Bede, noted that it was 'permissible' (in what sense? legal? moral?) to delay payment of a physician's fees for up to a year. In the 12th century papal edicts were issued, intended to prevent monkish physicians from charging any fee.

The civilised solution to this dilemma is to arrange for doctors to be paid by some means other than charging fees to their patients. Armies have almost always provided free medical services for their personnel. In the later Roman Empire most municipalities employed physicians to look after the poor. Gladiators, too, got free medical care. Grand feudal households often included a physician who was expected to give free advice to its poorer members. The same principle informed most Communist regimes and, more successfully, the British National Health Service, which gave remarkable value for money before it was undermined by the introduction of a management culture in the 1980s.

It's a tribute in part to the resilience of ideas, even when they seem clearly to be past their sell-by dates, that the National Health Service is rapidly unravelling and private medicine re-establishing itself. The idea of a physician's right to a fee from his patient is still at the core of the Doctor, despite laudable efforts to detach it and the hostility felt for it by a proportion of individual doctors. Nevertheless there are built-in safety mechanisms. Intellectual honesty, as well as esoteric learning, is at the core of the Doctor and tends to counteract fee-grabbing quackery. This circumstance of course produces its own strains, which were particularly acute in the days when evidence of their own efficacy was hard for doctors to come by and they relied on knowing that they must be right because they were following precepts generally agreed to be so.

Two further important strands that went into the Doctor came from very different sources: one derived from the Priest and the other from the Nature Magician or Herbalist. In ancient Mesopotamia, physicianly functions were originally subdivided into three, although the first group was later taken over by the other two. The Baru foretold the outcome of illness; the Ashipu drove out evil spirits responsible for disease; the Asu were general practitioners using herbs as their principal therapy. In Greece priestly tradition was represented by the Aesculapiads, who set up temples for diagnosis and healing. The Hippocratic tradition, which was to some extent opposed to temple medicine, was more naturalistic and 'scientific'. Although Hippocratic medicine appeared to win the conflict between the two traditions, it would be more true to say that the priestly one was subsumed rather than completely demolished and is liable to reappear whenever conditions are right. In the Middle Ages, for instance, monks, whose primary functions were religious even though few were priests, were probably better physicians on average than the lay leeches. More important, ideas about the role of God and sin in disease are always liable to be resurrected, even in periods like the present when naturalistic approaches to illness seem to have carried all before them. Some physicians, as well as some sectors of the general public, regarded AIDS when it first appeared as a judgement on immorality. Thankfully, saner attitudes have since prevailed.

The Doctor is too complex an idea to be capable of full manifestation in any one person. No-one can be a priest-logician-healer-herbalist-astrologer-alchemist-businessman at once. Thus individual doctors have tended to fall into sub-categories. Today's super-specialisation was foreshadowed by the Egyptians, who had doctors with titles like 'Keeper of the Anus'. Ophthalmology was a particularly important specialty among them. The commonest subdivision, however, has been into physicians and surgeons. The eighth item in the Hippocratic oath is: 'I will not cut those who are suffering from the stone, but I will leave them to men who practise such operations'.

Surgeons have generally been of lower status than physicians, though in India the surgeon was invariably a member of a high caste. Maybe an Indian influence, transmitted through the empire, was partly responsible for the unhistorical, and perhaps temporary, high status of surgeons in the West since the mid-19th century. The West also added the Apothecary to the sub-types of Doctor for a time. He dealt mainly with the lower social orders and had a treatment role as well as a dispensing role, so is best regarded as a re-expression of the old line of herbalists who had never been wholly converted into Hippocratic physicians. Interestingly, apothecaries, having been renamed pharmacists and demoted to being just dispensers of drugs early in the 20th century, are now beginning to re-emerge in an officially sanctioned advisory role.

Modern specialists, although all now receiving the same basic training, also tend to represent different aspects of the idea. Psychotherapists, for instance, especially if interested in dream analysis, have much in common with the Aesculapiads. An endocrinologist would be likely to feel quite at home with a 15th century medical graduate of Bologna or Paris. General practitioners, on the other hand, would probably prefer the company of leeches. Some private plastic surgeons, one might argue, could readily change places with Chaucer's physician.

Perhaps it was complexity that allowed stray concepts to attach themselves to the Doctor and hang on for centuries in a manner not seen in simpler entities (the Dance, the Saint), which show a more definite distinction between core and associated concepts. Galen, the best-known physician of the later Roman Empire, was the most prolific source of such notions. He is said to have written over five hundred books, fewer than a hundred of which survive. All contain misleading ideas which retarded the progress of medicine for some 1400 years.

The position of an American physician now is remarkably similar to that of the Egyptian doctor two thousand years ago. One Diodorus Siculus wrote about the latter, 'If the doctor had observed the rules of the so-called sacred books and had acted in conformity with them, but... did not succeed in saving the patient's life, he was free from blame. But if he acted contrary to the precepts he risked his life'. Substitute 'livelihood' for 'life' at the end and the two situations are identical – remarkable that an idea about the circumstances in which doctors are liable should have survived unchanged over so long a time.

Other concepts were less securely attached to the Doctor, but nevertheless lasted almost as long: the idea that effective medicine should taste nasty for instance. This is still alive, though moribund, and derives from a belief belonging to the priestly aspect of the archetype: namely that evil spirits causing disease can be banished by utterly revolting substances. The relish for bleeding patients, which led to doctors being called leeches right up to the 19th century, is often thought to have derived from the Greek notion that illness is due to imbalance of the four 'humours', one of which is blood. This belief may have contributed, but the main impetus came from Erasistratus, an Alexandrian born around 300 BC, who thought little of the humoral theory but taught that most diseases are due to overproduction of blood by the heart.

Like other cognitive objects, the Doctor can sometimes seem to have a life of its own, which can undermine the welfare of individual doctors. Their high suicide and alcoholism rates may be partially attributable to such conflicts between the idea and the person, or between incompatible elements of the idea[5]. An example, which is still operative to an extent and must have been particularly acute in the 18th century, is conflict between demands made by the requirement for erudition and other demands, such as those for intellectual honesty and compassion. Eighteenth century treatments were mostly very unpleasant and, in general, obviously did more harm than good – at least it was obvious to cartoonists and satirical writers. Yet the classical authorities that every good practitioner had to know about said these treatments were correct. The psychological stresses on doctors must have been huge. Even when a treatment probably was beneficial or life-saving there were still strains. A surgeon of the time, Astley Cooper, wrote of a relative in the same line:

> My uncle was a man of great feeling – too much so to be a surgeon. He was going to amputate a man's leg... when the poor fellow... jumped off the table and bolted; seeing which, instead of following the man, and attempting to persuade him to submit to the evil which circumstances rendered necessary, [he] turned round and said... 'By God! I'm glad he's gone'.

The sufferings of 18th century physicians were of course in no way commensurate with those that they inflicted on their patients. The point is that a 'good' idea like the Doctor follows its own laws, which may harm those who embody it as well as others who encounter it, contrary to the wishes of all concerned. Anyone who doubts this should recall the curious fact that the use of poppy, alcohol and mandrake root to diminish pain was well known to mediaeval surgeons, but only alcohol was used at all regularly for this purpose 400 years later. Even if the feelings of patients were considered irrelevant, the sensibility shown by Astley Cooper's uncle was unlikely to have been particularly rare; partially anaesthetising the patient would certainly have benefited him. Similarly, the role of fresh vegetables in preventing scurvy was familiar to Elizabethan and Restoration seafarers, but was not 'discovered' by naval physicians until the 18th century, and even then was often ignored for another fifty years. The backsliding happened, not through lack of practical knowledge, but because the uses of analgesics or greens for these respective purposes did not feature in learned texts approved by Renaissance physicians. Sir Gilbert Blane, physician to Admiral Rodney's fleet in 1780, was altogether exceptional when he 'took note of the common experience of seamen for generations'. 'It was a radical novelty that a physician should prefer folk wisdom to classical learning...', says Rodgers in his masterly telling of British naval history. Blane encouraged crews to eat fruit and fresh vegetables, and keep themselves clean (typhus fever transmitted by body lice was often even more of a scourge than scurvy in fleets), thereby helping to make the fleet probably the healthiest body of British subjects anywhere in the world at the time.

It is possible to conclude that the Doctor story is so large and complex that it may well be chaotic in the same way as the weather. Like all complex dynamic systems, it will harbour what are called 'attractors'. These are the mathematical objects that describe how a complex system will behave and show in what final state or states it will settle down. The attractors in the

Doctor system can be envisaged as corresponding to the various sub-types of Doctor and their differing attributes. If the system is in fact chaotic, very small differences in initial conditions will send individual doctors unpredictably down wildly different career paths; paths that they might well not have wished to choose. They would perhaps find themselves in much the same unenviable predicament as the lovers in Dante's *Inferno*, constantly blown from here to there by swirling gusts of wind.

Hospitals and asylums

These are wonderful places, with dedicated staff, to which the sick can go for care, cure and comfort. Or are they? Aesculapian temples of healing, the first Greek hospitals, certainly sound pretty marvellous. If you were sick, you could go to the temple for a night or two. There priests would purify you with ablutions and sacrifices, after which you would be led to a sacred area and allowed to sleep. If fortunate, the God would appear to you in a dream and you would be cured. Even in the absence of immediate cure, priests could advise you on the meaning of what you had dreamt. In the larger centres like Epidaurus, baths, sports pavilions and even a theatre were also laid on in the interests, one assumes, of preventive hygiene, both physical and mental. They seem to have perfectly exemplified a Thatcherite concept of what hospitals geared to customer preference should be. Yet they lost out in competition with individual treatment from Hippocratic physicians. Maybe the God came too infrequently, or the financial demands of the priests became in time too exigent. All the same, the idea of the temple or shrine of healing never wholly died and has contributed something to the idea of the Hospital.

The first Roman hospital was a very basic affair. It was situated on an island in the Tiber (there is still a hospital on the same site) and catered for old or ailing slaves who were dumped there when they became surplus to domestic requirements. The social dumping ground continues, of course, to play a large part in ideas about hospital functions, albeit a mainly tacit one nowadays. A more constructive Roman development related to the organisation of their legionary and field hospitals, which provided excellent hygiene and basic care. Interestingly, legionary physicians were not officers, though they did get special rates of pay and, unlike other soldiers, were allowed wives. Army hospitals were under the command of medically untutored people and were apparently all the better for it. Florence Nightingale, of course, taught a similar lesson. Her reorganisation of the military hospital at Scutari during the Crimean war was achieved against the opposition of some doctors, but proved a useful model for civilian hospitals of the period.

Mediaeval hospitals were at first of two main sorts: monkish infirmaries and leper colonies. It was part of the rule of St Benedict, followed by all early Western monasteries except those belonging to the Celtic church, that monks should care for the sick. This was usually interpreted to mean sick lay people as well as fellow monastics, and every monastery had its infirmary which was often airy, spacious, well-heated and provided with running water. Most also had gardens attached, giving a source of medicinal as well as culinary herbs. As populations increased, demands on such infirmaries grew and a need for specialist institutions was perceived.

The first specialist hospital, as distinct from almshouse, in England was probably St Peter's in York, founded in 936. The Hotel Dieu in Paris was founded by Archbishop Landry in 660, but this may originally have been more of an almshouse or social dumping ground than an equivalent of the monastic infirmary. St Bartholomew's, the first London hospital, was established in 1123 and has flourished ever since in spite of the vicissitudes of government policy. St Peter's, York, got into difficulties and had to be re-founded in 1135, under the name of St Leonard's, as a semi-monastic, Augustinian organisation. It quickly became one of the largest hospitals in the country, catering for over 200 patients. A special nurse was employed to look after sick children and two cows were reserved to keep them in milk. Because of the monastic connection, however, it eventually succumbed to the onslaughts of Henry VIII. The scale on which places were provided during this period was often remarkable. In 1300, for instance, there were about a thousand beds for 'the sick and needy' in Florence. This was rather more per head of population than most modern cities provide now for the same broad group (the Florentine ratio was something like 1:90, whereas the modern figure might typically be around 1:100). The figure is all the more surprising given that there was little need for geriatric care in the 14th century.

Leprosy probably reached its peak during the 12th century. The earliest recorded English leper hospitals opened around 1080, whereas there had been one in Paris in the 6th century. The first two in England were both near Canterbury; one was for lepers only, the other also took in 'the poor, infirm, lame and blind suffering from "several diseases"'. A century later, 20 leper hospitals existed in the county of Norfolk alone. An estimated 166 hospitals were founded in England in the 12th century, of which 80 catered mainly for lepers. When the flood of leprosy began to recede in the 13th century, some of these hospitals fell into ruin but others survived by caring for a range of chronically ill people, or for the indigent or aged. It's evident that there is a long tradition in the history of the Hospital of providing asylum and care for society's rejects as well as cure for respectable people.

The various western hospital functions got rather confused as well as glamorised around this period as a consequence of activities of the Knights of St John, or Knights Hospitaller. When crusaders unexpectedly conquered the Holy Land floods of pilgrims followed in their wake. Two principal military organisations were set up to help preserve the conquests and to protect and cater for pilgrims; these were the Templars, who specialised in financial services, and the Hospitallers, who initially concerned themselves with accommodation and almshouses for travellers, but later expanded into providing medical services too. They developed an excellent network of hospitals, in the crusader states and along the pilgrim routes, that remained prestigious for centuries. The Templars were destroyed after the Kingdom of Jerusalem was lost, whereas the Hospitallers went on to conduct heroic rearguard actions against the Turks. Their defence of Rhodes against Suleiman ('the Magnificent' or 'Lawgiver') and, later in the 16th century, a successful and even more heroic defence of Malta, are epic tales of courage and dedication which stirred many a heart. There is still an idea around that all hospital staff should be like Knights of St John, as evinced by the 1960s TV show *Dr Kildare* and its numerous successors.

In the 19th century, mental illness came to fill the role that had been leprosy's in the 12th. Lots of patients came to notice who were frightening, in a shockingly distressed condition, and above all difficult to treat. The solutions adopted were very similar in outline. There were precise regulations governing the attribution of leprosy or mental illness to a person, accompanied by a religious ceremony in the case of lepers and a legal one in the case of the mentally ill. By the mid-20th century the ratio of provision of mental hospital to ordinary hospital beds was almost identical to that of leper versus other hospital places in the 12th. Indeed in some countries, mentally ill patients were placed in abandoned plague hospitals or leprosaria. This occurred for example in 18th century Denmark and in Zambia in 1969[6].

The oldest UK mental hospital was Bedlam in London, which derived its name from the order of St Mary of Bethlehem. It was used for the reception of cases of acute mental disorder from 1377 onwards. As an institution, it had all the hallmarks of advanced decay throughout the 17th and 18th centuries. Public knowledge of this, together with scandals in private madhouses and awareness that large numbers of insane people were held in appalling conditions in almshouses and workhouses, led to a 'something must be done' mood by the end of the 18th century. Precisely what should be done was not clear at first, but one way forward was soon indicated by a wealthy Quaker family in York, the Tukes. They set up an institution which opened in 1797 called 'The Retreat', where seven staff looked after 30 patients in an atmosphere as much like an ordinary home as possible. Though much discussed and held up as an ideal, this approach was not easy to adopt on the large scale that was needed, and a rather different one took over.

There's a delightful essay by Christine Stevenson, writing in 1988, on how the new policy manifested itself in Denmark, a country which had no significant indigenous tradition of caring for the mentally ill other than the provision of lock-ups called *daarekister* (meaning 'mad-box') for the violently insane. Asylum in the countryside was the thing:

> In the bosom of country life St. Hans's [the old Copenhagen mental hospital's] sickly inmates would emulate the pure and innocent customs of the locals, and fall into natural rhythms of sound sleeping and healthy eating.... The Hogarthian vignette of urban life with its madness, venereal disease, and alcoholism, would be replaced by a picturesque landscape.

This attitude, though taking more sophisticated forms in other countries, fitted in with the whole Romantic groundswell of the times took hold. The Victorian asylum was born.

The dream soon turned sour. In the volume containing Stevenson's essay, the medical historian William Bynum and colleagues wrote:

> Overall, the asylum took on for a time a status as panacea equivalent to the steam engine, the rights of man or the spread of universal knowledge... The aspiration, widely entertained in 1800 of curing a relatively small proportion of lunatics, seemed by 1900 in danger of revealing the fundamental craziness of the human mind itself.

By 1990 it had all ended in tears, with the frenetic destruction of as many asylums as possible and their attempted replacement by pallid imitations of Mr Tuke's Retreat, many of which have more in common with an 18th century almshouse than with the actual Retreat.

The idea of the Hospital is thus quite complex and by no means all benign. Concepts concerning healing are certainly at its core and are responsible for much of the value of the modern general hospital. These ideas can cause problems though, when, as in the case of the Victorian mental hospital, they raise expectations that cannot be met. There's also the idea about staff being knights in shining armour somewhere near the core, which can do marvels for morale in the right circumstances, but also has its downside.

More obviously problematic ideas have to do with the 'leprosarium' role of hospitals; i.e. their functions as providers of care and, unfortunately, as oubliettes for frightening, contagious or otherwise disturbing people. Particularly when cures are unavailable or seen to be ineffective, institutions dealing with patients of this type readily become very nasty places indeed for a whole range of reasons, the principal of these being 'out of sight, out of mind' social attitudes, which naturally entail a reluctance to devote resources.

What happened to countless women in maternity hospitals, over a period of around 150 years up to the mid-19th century, is a good example of how ideas can do great harm, contrary to the wishes of everyone concerned. The notion that people needing medical attention should be in hospital[7] got women into maternity hospitals, and the idea that doctors know everything worth knowing allowed them to infect the women with puerperal fever. After all, Galen had said nothing about fever being transmissible by dirty hands. Of one unexceptional group of 65 women admitted to the Westminster lying-in hospital in the 18th century, 14 died of puerperal fever. In the following century an obstetrician in Vienna noted that 18% of mothers died who were admitted to a ward where doctors worked, but only 3% of those in a different ward with midwives but no doctors. He got the sack for his pains, and the massacre continued for another decade or two. A 20% death rate in maternity hospitals where doctors worked may have been typical, the baseline death rate from childbirth at the time being around 2%. The mean population of developed parts of Europe, with access to hospitals, over the whole period was around 90 million with an average lifespan, if neonatal deaths are included, of something like 30 years. This gives an annual birth rate of roughly 3 million. Suppose only 1% of those births were 'aided' by doctors in hospitals: the excess deaths of mothers caused by them would have been 5400 per year, or nearly a million altogether.

The Hospital too, then, is a complex idea that seems to contain a range of attractors. At present, the popular picture of them most resembles the monkish infirmary. Problems arise because it is hard to find people as dedicated as monks to run them. Attempts to overcome the problem have resulted in the incorporation of elements of the Roman legionary hospital, but have met with limited success, or outright failure. The 'asylum' function is currently in great disarray, and causes difficulties all round when attempts are made to amalgamate it with other types of hospital. Meanwhile, Aesculapian temples still manifest, mainly in the form of health spas. Although the idea of the Hospital does not appear to influence people's individual choices in any direct and obvious way, it certainly can affect how health service workers and the sick view the situations in which they find themselves, sometimes propelling them in directions that they might never willingly have chosen. The current confusion about the functions of mental health services or, to a lesser extent, geriatric services, shows how people can be swept along by tides of fashionable enthusiasm which appear to follow their own rhythms and laws.

Y

The examples given over the last four chapters suggest that cognitive objects can affect people in many ways. Some – the Noble Roman, the Dance, Fatigue, the Saint – seem able to take individuals over, either temporarily or for long periods; even for a lifetime in the case of the Saint. In the Dance and Fatigue, the takeover is involuntary and experienced as being like an illness. Where people chose their idea, as with the Saint and, to a lesser extent, the Noble Roman, they essentially *became* the object. Other ideas – the Millennium, the Doctor, the Hospital – act more like constraints on free will. They determine the range of choices available or alter the probabilities that particular choices will be made. Also, some cognitive objects may be so complex that their dynamics are chaotic. In a chaotic system, a person caught up in it could make a choice expected to lead directly to some particular outcome, but might then discover that the choice had in fact put them on a trajectory leading far away from what they intended. In these circumstances any feeling of being in control of one's life would be illusory for all practical purposes.

Perhaps the most remarkable thing about the objects described is that they have deep roots, sometimes going back to classical times or further. I've only described a small, idiosyncratic selection of objects and cannot prove that all, or even most, similar objects have equally long histories. Nevertheless, the importance we give to notions of culture and cultural continuity does suggest that such examples may be fairly typical. They can be thought of as stories unfolding over time, in much the same way as individual people can be viewed as tales (see Chapter 6). But the story of the object often 'writes' that of the individual, it seems. Individual 'stories' must often conform to the structure of much larger and long-lasting tales. Does this leave any room whatsoever for individual free will, or is all culturally determined? Let's return to Susan to find out.

Chapter 11

SUSAN'S TALE (PART II)

Susan and family returned to England about six months after her encounter with Professor Libet. It all seemed very gray, damp and constricted after California, but she wasn't too bothered. Her A-level results were good, and she would be starting at university soon. In fact, she had managed to get a place in medical school, which was generally agreed to be the best career option for anyone with science passes. Also, the choice fitted nicely with her concept of herself as 'good'. In the meantime, she took a temporary job in a pharmacy. It was a bit boring at times, but some of the other girls were nice and of course the money came in useful.

Medical school proved something of a marathon. Everything worked on 'continuous assessment', so there was no chance of skipping courses and catching up later. Nor could you readily take time out to follow your own interests. There were endless facts and theories to be learned, and an almost overwhelming mix of new experiences. People dying, people crying, angry people, foolish people, crazy ones too. 'And that's only the staff…', she sometimes thought wryly. In the fourth year, you were supposed to spend six months on a research project of your choice, but little 'choice' came into it, as the time was far too short to plan and carry out anything original. So it was a matter of joining an existing project or doing something, inevitably a bit half-baked, suggested by one of the medical school staff. There was really no time to digest all this new stuff, especially as the explicit learning was only the half of it. There was an equally overwhelming amount of implicit learning to do with skills of all sorts, from how to take a blood sample to how to talk to a bereaved relative or calm a difficult drunk. Susan absorbed these skills mainly through being there when others were doing it. She had very little conscious choice in the matter.

She passed her finals of course. She was not the sort of person to fail, or fall by the wayside. Her sister, Diana, was by this time a buyer for a chain store, with a flat of her own and a nice company car. When they went out for a celebratory drink together, Diana paid and Susan had one of her momentary twinges of envy. She was quite heavily in debt. But she still felt that she occupied the moral high ground. 'Doctors are useful and important... right? Who would miss a buyer?' Then she remembered the tale from *The Hitchhiker's Guide to the Galaxy*, which had been on the radio a couple of years previously. The one about the civilisation that died of a plague contracted from telephone handsets, when it lost all its hairdressers, telephone handset cleaners and other 'useless' people. Somehow, even when she did not mean to, Diana was always able to make Susan feel a bit ashamed of herself.

The next step was house jobs, or interning. You had to do two of these for six months each, one in medicine and one in surgery, before getting to be fully qualified. Looking back afterwards, Susan could never remember much about the jobs. They passed in a haze of fatigue. She was working a minimum of ten hours a day during the week, plus two or sometimes three nights on call when you could be guaranteed to be up half the night, plus every other weekend. The best that could be said for this time was that she survived it. Although she did not fully realise it, she was thereafter a very different person from the one who had entered medical school. She was full of ideas and attitudes, skills and habits that had got into her without going through the filter of her own conscious choice. She had of course chosen to be there and to take notice, but that was about all. For almost the first time in her life since infancy, when her 'choices' were often due to her brain's hard-wiring, her conscious self had had little or no responsibility for the changes that had occurred in her. In fact she had often felt during those years a bit like a passenger in a car, able to watch the scenery pass by but with no control over route or destination. It was only towards the end of the second job that she arrived at the first major choice, other than deciding when to dump a boyfriend, that she had been able to make since leaving the pharmacy six years previously. She decided to apply for a post as a trainee neurologist. She was never quite sure why she opted for this, out of all the possible options. It was a subject that appealed to her somehow, and she wanted to stay in hospitals rather than go into general practice. It just seemed the right thing to do.

That's why Susan is now sitting in a neurology clinic awaiting her next patient. This is a lady, known to the reception staff as Weary Wendy, who demands a wheelchair and someone to push it whenever she arrives at the hospital. Nothing has ever been found to account for her inability to walk across the waiting room. Despite her frequent claims of being worse than ever, she never looks particularly ill. On the other hand, she never improves either. Several attempts have been made to discharge her back to the care of her GP but she always bounces back after a month or two, with a new referral letter from some doctor somewhere. A nurse wheels her in and leaves her with Susan.

'Hello Wendy', says Susan, 'Nice to see you again. How are you today?'

The question is a mistake. Wendy tells Susan exactly how she has been today. How she felt so awful this morning that she could not get up to make a cup of tea till ten o'clock. How the effort of making the tea exhausted her so that she had to go back to bed and could not summon up the energy even to wash until mid-day. Then she had to get ready to come to the clinic, so she knows that she will be totally exhausted tomorrow and not able to do anything at all. She is speaking, Susan observes, in a lively, emphatic voice and shows no signs whatsoever of running out of steam while recounting her woes. Clearly the diagnosis of ME in the notes is correct. Wendy's symptoms are real enough from her own point of view, but people whose fatigue is due to definite neurological disease sooner or later also show signs of it that others can spot.

Susan thumbs through the notes, wondering what to say or to suggest next. Wendy's monologue is still continuing, though Susan has stopped listening. Then something catches her attention: 'I read about that wonderful magnetic therapy you can get in Upchester', Wendy is saying. 'You have to stay in the clinic for a month, but they get very good results with cases like mine. Do you think I should go there?'

'Yes, of course you should go', Susan longs to say, 'Anything to get you out of my clinic'. But her professional training clicks in to prevent her. Doctors don't approve of quack therapies. Doctors know what is causing diseases and, if they don't, they have a responsibility to find out. Doctors care for their patients.

What she actually says is, 'Well, it would cost you a lot of money, you know, and there's no evidence that magnetic therapy does work. I think we should soldier on here before you put yourself to a lot of expense. Now we have ruled out some of the viruses that might have caused your weakness, but there have very recently been reports that it can be due to a virus that vets can catch from horses. I remember you told me you used to like riding, so I think we should arrange some more tests to exclude that. Here's what we'll need to do...'.

Susan can hear herself speaking, but she does not really seem to be saying what *she* would have chosen to say. It is more as if some automatic 'doctor programme' is using her voice. This does not worry her, as it is such a familiar feeling these days. With the best of professional motives, Susan has just arranged for yet more tests that will lead nowhere and has reinforced Wendy's erroneous ideas about the nature of her condition. Yet it was not Susan who acted so much as the Doctor within. The two of them are growing closer, but are still to some extent distinguishable.

A month or so later, though, something happens which really does need Susan's personal choice, and has a profound effect on her future. Her boss, Tom, has a bout of flu and she is doing his clinic for him. Tom is a well-known, rather flamboyant neurologist with a taste for expensive cars. He has recently acquired a new four-wheel drive roadster that cannot normally be bought in England. Tom is especially well known for conducting trials of new drug treatments for Parkinsonism, but just lately he has also been looking into the effects of interferon on multiple sclerosis. The drug has to be given by injection, and careful assessments of any effects need to be made and recorded every day. One of the patients on this trial is booked into the clinic, and his research notes are there in addition to the standard health service folder.

Susan already knows what needs to be done for this patient, so she examines him and makes some notes in the health service folder. Then she opens the research folder, and looks for the appropriate questionnaires to fill in. Ooops! They are already there, filled in, dated today, signed by Tom. Her mind goes blank. She looks again. The completed forms are still there. She takes some blank questionnaires and fills in the results of her own assessment. They are less favourable than the spurious set. Having done this she gets on with the rest of the clinic somehow, though her mind is still in a whirl. By the time she's finished, it is nearly seven o'clock and all the reception staff have gone home. She re-opens the offending research folder and looks at it. There's no doubt. Tom has been falsifying results. She photocopies the offending sheets and tucks them in her handbag, then goes to Tom's office to look for other research folders. They are in a filing cabinet, but it is locked and there is no key. She goes home to her flat.

It's not a good night. Thoughts chase one another through her brain. 'He's my boss, he can't have done it'. 'What should I do? Ask him? Tell someone?' 'He's not going to like it'. 'I want to stay in this hospital'. 'But this is *research*, you can't just make up results whenever you feel like it. It's dishonest'. 'It's more than dishonest, it's dangerous. People are going to be *treated* because of cooked results'. 'But he's a consultant. They don't do things like that – do they?'[1]

Susan phones Tom the next day, still not having made up her mind what to do. She asks how he is, and tells him about some of the patients that she has been dealing with for him. Then she says, in as neutral a tone of voice as she can manage, 'And what would you like me to do with the trial forms you filled in?' There's a long pause. 'Oh those', says Tom, 'Throw them away of course. They are just a little game I play with myself. To see if I can predict how the real results will turn out. I fill them in in advance and then compare them with the real ones. And give myself a big pat on the back if the two are not out by more than ten points'. 'I see', says Susan, and rings off. She doesn't 'see', of course. She is still in just as much of a whirl as ever. 'Could what he said be true?' 'It's really not very likely'. 'Anyway, even if it is true, he's risking biasing the results because he'll want them to be the same as his so-called prediction'. 'He's not that silly. It can't be true – can it?'

She goes on in this state for several days. Then, quite suddenly, just before lunch, her mind goes still and quiet. It's as if she is standing at a fork in a road and is utterly free to take either of the two directions. She can keep quiet and get on with her life, or she can try to bring the fraud to light. She understands that the second course is going to be a difficult one, but there's no feeling of pressure to go in one direction rather than the other. All the turmoil has vanished somehow. It's simply up to her to decide which way to go. She stays poised between options for a timeless moment, then, 'It was wrong', she says to herself. 'I shall tell someone'.

The moment she's made the decision, all the uncertainties come rushing back. 'Is it the right decision?' 'Is it fair on Tom?' 'What about my career?' 'Was it really fraud?' All the same, she's not going to change her decision and the only real question now is how to carry it out. 'No good going to the chairman of the division of neurology – that's Tom'. 'The hospital managers? Don't trust them'. 'The press? There's not enough evidence, just a few photocopies that I might have faked because of some grievance'. In the end, she goes to the chairman of the hospital ethics committee that has to approve all research, including Tom's drug trials. He hears her out, looks at her photocopies, and tells her to 'leave it with him'. She goes back to work, feeling both relieved and dissatisfied at the same time. She doesn't have time to brood on this, though. She is too busy.

Nothing happens for a couple of months. She keeps an eye, when she can, on Tom's trial and the patients coming for assessment. So far as she can tell, he is doing everything just as he should. Then he calls her to his office one day. 'I'm sorry we don't get on as well as I'd have liked', he says, without preamble. 'I see you are coming to the end of this registrar post. I'd just like you to know that we have a lot of excellent applicants for the Senior Registrar job. I expect you'll want to apply, but the competition really is very stiff. You ought to consider putting in some applications elsewhere as well.' There's really nothing Susan can say or do. She gets the message. The professional old boy network has smoothed everything over, to the satisfaction of nearly everyone concerned. Irritants like her, right or wrong, are to be smoothly sent on their way. As it eventually turned out this was not such a bad thing in the long run, not for Susan at least. After a whole series of adventures and vicissitudes, she finally settled down as a highly respected professor of neurology – at a medical school somewhere in Australia.

She often looked back at her moment of decision, in later years, wondering how life would have turned out had she taken the other path. Being a neurologist and an academic, she asked herself, too, who or what had made the choice and why it had seemed so utterly free at the time. None of her other choices had ever seemed both so significant and so totally uninfluenced by anything other than her own decision. But who was 'her'? There were a whole range of possible answers to that question, she thought. Maybe 'she' was her body, plus the neural systems that report on her body and construct emotions. That was what Antonio Damasio, a fellow neurologist, was writing from his American base. Or maybe she was simply the sum total of her conscious experiences throughout her life; quite a nice idea in abstract, perhaps, but not one that could be applied to a decision maker as most of those experiences were long gone, leaving not a trace behind.

'It's the traces that matter', she concluded. 'The "I" responsible for decisions must be the sum of all the habits and memories that I've accumulated. Not quite the same as the "I" that I experience, which is a lot more like Damasio's picture'. 'After all', she continued, sounding to herself a bit like Alice in Wonderland trying to reason something out, 'I can't experience all of my memories all at once, because working memory simply hasn't got space for them all. And many of my habits are always unconscious, so choice-making "I" and experiential "I" must be different. They may overlap to some extent, but they are still not the same. But of course I already knew that a long time ago, thanks to Ben Libet'.

The question of why the choice had seemed so very free proved harder to answer, though. The answer that she liked best in the end went something like this: all 'free' choices have elements of bias in them due to current circumstances or your personal history. If you are tired, for instance, you're more likely to opt for a hot bath than to go out jogging. If you are feeling fresh, the reverse might apply. If you've just read a book on 'jogging is good for you', maybe you are more likely still to choose to go jogging. The more complex the issues involved, the more subtle and various will be the factors influencing your decision. But suppose all the opposing pressures, to decide one thing rather than the other, just happen to cancel one another out. Then only conscious choice will be left, and some rather abstract idea of 'I', like the essence of the sum of all one's memories, can be responsible for that choice.

'When I was deciding what to do there were so many factors involved', she thought to herself. 'Straightforward fear, self-interest, all my training in teamwork and respect for seniors were urging me to keep quiet. On the other side there was anger that anyone could do that, my knowledge of how research should be done and the consequences of not doing it properly, and all sorts of ideas about what's right and what's fair. The two sets of influences must have been evenly balanced and allowed some essence of "I" to show itself for once'.

The telephone did not ring to interrupt her thoughts for once, so she went on ruminating a little longer. 'Of course someone could say that my decision was nothing but an outcome of the deterministic or chance activity of some set of decision-making neurons – probably ones sitting in my frontal lobes somewhere. And that would be perfectly true, in a sense. All that I do or experience can be viewed as a consequence of special brain activity of some sort, and therefore wholly subject to physical law or to chance. But somehow all that's gone on in my life, especially the stuff I've been conscious of, has built itself up into a collection of ongoing stories that mean something. For instance there is a story that's been in my mind for as long as I can remember, about me being a fairly honest person. And another that's been there since I was a teenager, to do with being someone who cares about medical research. It was the stories that were *really* responsible for my decision. No doubt they are encoded in neurons, which are just physical things behaving physically, but it was not the neurons as such that determined my decision – it was the stories that did, and they *are* me'.

And that's where we shall leave Susan. She had a satisfactory life on the whole, and was mostly happy. Even when she was quite old, you could still see elements of both *Doctor Who*'s eponymous hero and his assistants in her – if you looked hard enough.

Chapter 12
STORIES AND THEIR PEOPLE

Great storytellers have often been ambivalent as to whether it is events or people that determine the course of their tales. Homer frequently regards his heroes as mere playthings of the Gods, but sometimes implies that the heroes' decisions are their own and make a difference. In *War and Peace*, Tolstoy seems to say that extraordinary men and their actions are no more than products of extraordinary times, though in *Anna Karenina* individual decisions determine the course of events. In that strange epic *Lord of the Rings*, which has so infiltrated the hearts and minds of two or more generations, there is a constant tension between the decisions that Frodo and others make and the impersonal sweep of the history of the Ring. Wilkie Collins was not in quite the same league, but stated outright, in the introduction to *The Moonstone*, that he was writing a book about the effect of events on people after his success with another (*The Woman in White*) about the effect of people on events.

Working out the relationship between individuals and their circumstances is like looking at a Necker cube[1], or the vase/faces illusion: the configuration flips every now and again. Sometimes circumstances appear to control people, at other times the individuals seem to determine what happens. Both points of view are equally valid. What remains constant in a Necker cube are the lines on the paper. 'Story lines', similarly, are embedded in both people and the march of events. Edwin Hutchins, an anthropologist based in San Diego, said something rather like this. He is deeply fond of both his profession and his hobby of small boat navigation. Reconciling the two, he made a study of navigators in the US Navy and was particularly impressed by finding that aspects of their conscious behaviour appear to reside in their instruments and group methodologies, not in their individual minds. However, navigators have of course themselves created the instruments and the methods over many centuries of effort, even though the experience of any particular navigator seems in part a creation of the instruments. What is the overall structure of story lines of this sort?

The descriptions of the 'cognitive objects' (i.e. story lines) in previous chapters were rather biased in the direction of them being in control of people. It was a perfectly valid point of view, adopted to emphasise the point that ideas *do* affect people independently of their own wishes or intensions. But it is equally true that people influence and rarely even create ideas. Some individual Roman no doubt either implicitly or explicitly decided that he and his associates should begin to behave like patricians, thus contributing to the birth of the Noble Roman. Some south Italian citizen must have been the first to confuse the allegedly poisonous lizard with the tarantula spider, so setting the scene for Tarantism to develop. Although the origins of the Millennium are lost in the mists of history, individuals like Lactantius put their own gloss on the tale and influenced its development. Neurasthenia would almost certainly have been discovered by someone else if Beard had never become a neurologist; nevertheless he was pivotal in its history. St Francis and St Ignatius put very different personal stamps on the nature of the Saint, and thus founded orders that have had quite different histories. St Benedict created, in some sense, the rule that later assisted the development of the Hospital.

The picture building up here is that, rather as Susan concluded in the previous chapter, choices that are felt as free can sometimes be made when an individual's collection of memories, habits and so forth (i.e. their own 'story') interacts with one or more of the ongoing stories in their environment. These 'external' stories form the skeleton of the culture, both social and material, that envelops us all. And a person's choices will actually *be* free to some degree, in the sense that the conscious part of the brain does have some autonomy in relation to the rest of the brain. The memories (from shortest to longest term), through which consciousness influences its own future and that of 'its' brain, are inevitably an idiosyncratic assemblage that can never be quite the same in any two people, not even identical twins brought up together. Therefore, when someone interacts with an ongoing story, the outcome is going to be unique to some degree. It may usually be quite similar to the outcome of another person's interaction, since people aren't necessarily different enough from one another to make a difference to what transpires. Similar outcomes will reinforce and help to perpetuate whatever story it is, but every now and again something unusual may occur, capable of altering the course of events and the nature of a story line.

As we have seen, individuals can internalise these external story lines. Such cognitive objects can become part of a person, or occasionally most of a person, as in Saints. We can thus envisage people as nodes in a vast, three-dimensional network of story lines; the dimensions in question are two of space, corresponding to how people are scattered over the surface of the world, and one of time. The set of lines entering each person or node is unique, but broadly similar to those penetrating most neighbouring nodes. A node may sometimes alter a line that passes through it. Very occasionally a node may originate one or more entirely new story lines, though altering an existing one is the more usual type of innovation.

One of the most interesting characteristics of the network is that it is more extensively linked in time than in space. Cultures tend to resist ideas, influences and indeed individuals from neighbouring cultures, particularly contemporaneous ones. The conscious attempts of the French to keep out English words, or of the English to keep at bay American spellings, are examples of this characteristic. However, influences extending over time are of great importance to continuity within a given culture, and are better able to penetrate neighbouring cultures[2]. They can seep in almost unnoticed, in the way that a liking for curries permeated British culture starting in the 18th century, and is now manifest in the Indian restaurants that proliferate in every town. As we have already seen, story lines rumble down the centuries and millennia, moulding the characters and behaviour of all they interact with.

At the basic level are the links associated with nuclear families, then extended families and neighbourhoods, then workplaces, towns and professions, then regions, languages and countries, then international organisations, religions and whole civilisations. On the largest scale is humanity in its entirety. Most nodes (i.e. individual people) have lots of relatively short-range connections and few longer ones; a few have a greater proportion of long-range links, either spatial or temporal or sometimes both. In fact the whole network was probably already quite like the World Wide Web, before ever the Internet was invented. Modern communications allow it to be more extensive than it was, and have speeded up some of the temporal links, but may not have affected the overall network structure, which could prove to be scale-free – that is, it may look much the same whether you just take a little chunk to examine, or a bigger one, or the whole thing.

The temporal connections tend to be more influential, as well as more extensive, than spatial ones. Newton, for instance, has almost certainly had a more profound influence on Western civilisation than his contemporary, Leibnitz, who in many ways was at least a match for Newton in the originality of his thinking. Newton was something of a recluse who published relatively little of his work, while Leibnitz was constantly on the move and a prolific writer. Newton's work was read by few of his contemporaries – with good reason; the *Principia* is virtually incomprehensible to anyone who is not a trained mathematician, and there were very few of those about in the 17th century. What made the difference was that Newton's ideas were better preserved over time, by the Royal Society and by people like Voltaire[3].

The smallest network scales are characterised by links to do with ideas such as, 'people in our family like pets' or 'we have tea at six o'clock' or 'our boys always get jobs in the docks'. Slightly larger ones have to do with notions such as 'we Yorkshire people are careful with money' or 'we all support Leeds United football club'. The ideas described in previous chapters have to do with still larger network scales – professions, nations, civilisations and religions. On the largest scale of all, that of our species and all that has happened to it so far, are the ideas Jung termed 'archetypes'. The concepts he wrote about included Mother, God(s) and Mandalas (i.e. symbols such as a cross or a lotus flower enclosed within a border, usually a circle).

The origins of meaning

The networks pictured here are of meaningful ideas, and they gain their meaning from their nodes – people. At the same time, each idea imparts meaning to the nodes it enters. In other words, there is another loop here, not of neural impulses this time, but of meaningful information. Like much of nature, the 'bootstrap principle' seems to be at work. Neither the nodes nor the lines connecting them are sufficient on their own to generate meaning; it emerges only in the context of the network as a whole.

Of course an isolated node probably could generate some sort of simple meaning. A newborn child washed up alone on a desert island, if she survived for long enough somehow, could learn that seeing a coconut tree means the possibility of getting a meal. That sort of meaning must presumably derive from an internal network, exchanging information between eyes, appetite centres, memories and so forth. But any more complex meaning has to be based on social, node-to-node, networks.

But where does meaning *really* have its home, someone might ask; is it in the people or the ideas connecting them? A Westerner's gut reaction would be, 'the people, of course'. A Buddhist might answer somewhat differently. The Dalai Lama put it this way:

> ... we might wonder what the mind is. If you ask them, most people respond by rubbing their heads and pointing to their brains. This is partially correct... the human mind does not have any existence independent of the human body... the human mind or consciousness... actually consists of a vast network of minds, some subtle, some coarse...

The 'network of minds' to which he refers thus appears to include both what we have called 'cognitive objects' (so also referring to the story lines of the network) and presumably neural modules like those discussed in Chapter 5. Perhaps, then, from a Buddhist perspective both people and stories are equally illusory. It's only the combination of the two that achieves more substantial meaning.

The truth is that neither lines nor nodes, separately, have much reality from the experiential point of view. Concepts cannot exist without people, but the mental life of people without concepts is impossible to imagine. All one can say is that it would be extraordinarily impoverished. Together lines and nodes generate the rich world of meaningful, experienced reality. This being so, it is understandable that meaning should have proved so intractable a concept from a mathematical point of view. The popular mathematician Keith Devlin has written a couple of very readable books describing how information theory is wholly inadequate when it comes to dealing with meaning. He advocates substituting entities called 'infons' for bits or bytes in order to build a theory. But his idea does not seem to have caught on, so maybe even infons aren't adequate for the purpose. The fact is that interactions between lines and nodes could not realistically be modelled by holding one or other constant in the way that mathematicians like to, because neither stays constant when they interact. Moreover, their interactions are probably non-linear, which would add to the mathematical difficulties. A possible way forward is offered by something called 'algebraic semantics', but this is currently quite an esoteric subject at only an early stage of development[4].

So although an important goal for any science is to develop appropriate mathematical models, and the network picture seems to invite this, there is probably no realistic short-term prospect of achieving any complete treatment that captures the quality of meaning and the structure of the network. However, a description of the network structure alone would be of great interest. Network theory is actually quite a vibrant field at present; a good deal of work has been done on the structure of the World Wide Web, for instance, and there is the famous 'six degrees of separation'[5] configuration that seems to approximately characterise many social networks. The brain itself has a similar structure, though individual neurons are mostly separated by fewer than six degrees. It might not be such a big step to describe World Wide Culture – the overall structure of the interactions that are relevant – even though there is no hope at present of describing how or why these interactions occur. And doing so could prove very worthwhile. After all, there is no knowing *how* or *why* things happen without first knowing precisely *what* is going on.

I've been claiming throughout that my account of free will is scientific. Perhaps I should have defended this claim before getting on to its possible future mathematisation – better late than never, though. Despite all the subsequent doubts about Karl Popper's description of science, it remains true that one of the best ways of distinguishing science from non-science is to see whether whatever is being said is based on refutable propositions. My account is certainly refutable in principle. The two 'axioms' on which it is based could, despite all current evidence to the contrary, turn out to have been wrong once we have a complete understanding of consciousness. It's possible that distinctive conscious states will be discovered that are not associated with distinctive neural states. Maybe consciousness will turn out to be a late arrival on the scene for some reason other than close entanglement with the memory process. The first finding would certainly throw a lot of doubt on the picture I've developed, while the second would totally undermine it. However, it is likely to be a very long time before we do have a complete understanding of experience, so considerations like these are about 'in principle' refutability only. Are there more accessible tests that could be applied?

Well, two of the main planks in my argument are to do with the relationship between memory and freedom. Memory frees us from the de la Mettrie machine, while the fact that long-term memories can be edited and modified contributes to freeing us from social determinism. Both of these propositions are testable. For ethical reasons, one could not abolish someone's memory and see if their freedom vanished along with it. On the other hand, it ought to be possible to compare the 'freedom' shown by Korsakov patients (see Chapter 3) with that of other people whose memories are OK but who find themselves in comparable situations. The Korsakov patients should have less freedom – as demonstrated by the range of choices that they are able to make, for instance. Maybe this is not a very interesting prediction, because it is so much in accord with common sense. If you can't remember that you've had rice pudding ten days in succession, you're surely more likely to opt for rice pudding on the eleventh day than is someone who *can* remember.

The predicted association between modifiability of long-term memory and freedom from social determinism is more interesting, though, because somewhat counter-intuitive. My claim is that people whose memories are more malleable should, other things being equal, be less prone to conditions like millenarianism (Chapter 8) or ME (Chapter 9) because they ought to be more able to modify 'cognitive objects' encountered in their cultures so as to fit in with the rest of their lives. Common sense might predict the opposite; namely that people whose memories are more malleable are likely to prove more susceptible to 'outside' cultural influences also. The problem with actually doing a study like this would lie in the 'other things being equal'. Except for the memory variable, subjects would need to be very similar indeed in all sorts of ways (e.g. age, sex, intelligence, personality traits, cultural background, cultural milieu) for any comparison to be worthwhile. All the same, difficulties like these are not insuperable, so the refutability of my account does not depend only on vague suppositions about our future understanding of consciousness. It is indeed scientific.

Y

We've now got about as far as it is possible at present to go in constructing an account of free will. At least we've come as far as it is possible to travel with any confidence, in the light of contemporary research. To recap, we've seen that thanks to its intimate relationship with the memory process, consciousness can to some extent determine its own future. It is not wholly at the mercy of unconscious neural mechanisms. We've found that the *feeling* of being in conscious control is rather separate from the actuality of conscious control, which mostly relates to far longer timescales than the feeling itself. Then we looked at how cognitive objects often play a central role in decision making, and may sometimes produce outcomes that bear little relationship to the conscious intentions of the people harbouring them, while at other times they can appear to take over people's consciousnesses altogether. Now we see that there are intimate reciprocal relationships between these objects and people as individuals, which nevertheless still leave a place for an influence of individual consciousness on what will transpire. Consciousness can, to some extent, select which cognitive objects it will incorporate, and it can sometimes modify the objects in accordance with its own preferences. Again, memory plays a crucial part here, both in housing the objects and allowing consciousness to work on them over the course of time. That long-term memories can be re-edited after recall (see Chapter 3) may be crucial in allowing consciousness to modify existing story lines.

There is a big 'but'. It is that this view of a partially autonomous consciousness has been reached through picturing it as a part of the physical world – as simply an aspect of brain function. Therefore, although it may seem free to itself, and indeed may actually sometimes *be* free in relation to the rest of the physical world, it must itself be bound by physical law, as is any other part of the physical world. Thus, in any ultimate sense, it cannot be regarded as free after all. It does, to some degree, have responsibility for itself and 'its' brain. It is captain of its own ship to some extent. Despite all this, it seems in the final analysis to be a captain whose behaviour must be considered just as deterministic as that of the autopilot. Many people would say that this conclusion has to be accepted. It is just part of having a grown-up attitude to the world. Conclusions deriving from sound science have to be accepted as definite. I disagree.

Some conclusions deriving from good science are indeed conclusive, but they are quite rare. The explanation of why mirrors reverse our image left-to-right, but not top-to-bottom, *is* one that's unlikely to be superseded (in case you have not come across it, see note 6. It is both ingenious and a little surprising). But most scientific accounts are tentative and liable to be improved upon. For more than a thousand years Aristotle's rationalisation held sway: 'earthy and watery things fall because they are attracted to the earth, while airy and fiery things rise because they are attracted to the sky or to the empyrean'. Then came Newton, and his: 'things fall because they possess mass and are subject to gravitational attraction to one another; anything that rises does so only because it is displaced upwards by other objects having greater mass'. Newton's account, while very different from the earlier one, retained a concept of 'attraction'. Now we have Einstein's: 'things fall because they have to follow an inertial path through a curved space–time unless acted upon by some external force; the space–time is curved by the presence of mass'. 'Attraction' has disappeared altogether, it appears. In fact, Einstein's explanation does include Newton's as a special case and 'attraction' retains a sort of shadowy implicit existence; but the new explanation has a very different feel. Einstein's version, too, will eventually be subsumed into or overturned by some yet more accurate description.

It is hugely probable that the account of free will I have given up to this point will turn out to be no more than a stepping stone to something better. It is based on concepts of matter and information that still conceal enormous mysteries, and it involves meaning – on which science has hardly begun to get a grip. It's worth remembering that Claude Shannon's definition of information, the one that scientists mostly use, proved so enormously useful precisely because he excluded any notion of meaning from his idea of what constitutes a 'bit'. If information is itself to a degree mysterious, then meaning is even more so. In consequence, any better model of freedom may look very different when we have it.

What sorts of development may one day lead to a future, more complete, description of free will? Maybe our present concept of 'determinism' will turn out to conceal deeper nuances, rather as gravitational attraction, which certainly seems to be a dominant force in all our lives, is actually a sort of illusion produced by the curvature of space–time. It is always risky to forecast future developments in science – one is far more likely to get them wrong than right. So here's a health warning:

Too much speculation can damage your rationality.

Those concerned for their safety should skip straight to the 'Epilogue'. Braver souls, follow me...

Chapter 13

OF LAWS, PHYSICS AND STORIES

The speculations described in this chapter are organised into an argument a bit like a tripod on three, somewhat wobbly, legs. The first and most unsteady of these concerns the role of causation in scientific explanations. Next is a brief look at what sort of thing matter might really be. Finally, I touch on two very unusual types of subjective experience. This could prove a bit of a bumpy ride for some tastes, but then... who expects second-guessing the future to be smooth or easy? You will need to abandon affection, if you have any, for the extreme reductionism that characterised some 20th century science. The new century focuses on complexity, emergence, the sheer wonder and richness of the world we inhabit.

Causation

Aristotle classified causes in a way still useful today. He said that there are four types. These can roughly be translated into more modern terms as:

- Material causes (the ball bounced because it is made of rubber)
- Efficient causes (the ball bounced because I threw it)
- Formal causes (the ball bounced because of the laws of dynamics)
- Final causes (the ball bounced so that the dog would go fetch it)

Final causes are rightly regarded with great suspicion when they crop up in scientific explanations (which they sometimes do, usually under the name of 'teleology'). There is no agreed means of handling them and they often turn out to have been extremely misleading. We shall have no truck with them here. David Hanke of Cambridge University, for instance, has described how teleological thinking in his field (plant biology) has badly distorted research priorities. People assume that the 'purpose' of plants is to be green-light processors, so they tend to ignore the non-green, etiolated state, which is actually in some respects the more complex and interesting of the two.

Material causes are important in research, but mainly as background factors, and are often referred to as ontology. Once you've agreed on the appropriate ontology for the piece of science you are dealing with, you can get on and look for the 'real' causes. Along with the majority of philosophers and neuroscientists nowadays, I have presupposed that the appropriate ontology for explaining free will is some form of materialism, so can we forget about causes of this type for the purposes of this chapter? As it happens, no.

Efficient causes are the meat of all scientific explanation of how one thing gives rise to another. When scientists, or indeed the rest of us, use the word 'cause' it usually means 'efficient cause'. They provide explanatory chains like: the window broke because the ball went through it because I made a bad hit with my bat.

Formal causes, on the other hand, are the physical laws. The window broke because of the law of conservation of energy. When the ball slowed down on hitting, it transferred more energy than the window could withstand. These causes constrain what types of efficient cause may exist and how they can operate. Windows and balls exist courtesy of the laws of quantum field theory (ultimately) and interact in ways determined by the laws of dynamics and solid state physics. One of the main goals of research is to discover and describe formal causes. Discovering a really important one is the quickest route to a Nobel prize.

But the boundary between efficient and formal causes is a lot fuzzier than is often supposed. Clearly there is a one-way flow of explanation only between, say, the law of conservation of momentum and the 'efficient cause' account of how a pool-table shot last Saturday night sent the ball into a pocket. In that case, we are talking about a local, specific event and a law that governs the behaviour of all matter everywhere. However there are so-called laws that could in principle, if not in practice, be deduced from more fundamental ones by looking at the relevant efficient causes; those of hydraulics for instance. These are in a sense the *product* of efficient causes, namely the interactions between water molecules (interactions which are themselves ultimately down to quantum theory). Then there are bylaws, if you will, like those of physiology. These are very obviously no more than emergent or derivative properties of the efficient causes operating in the blood vessels or lungs, say. Nevertheless they behave *as if* they govern the operation of the system in question.

Psychologists often moan that they do not have any laws to match physicists' discoveries. Maybe they are wrong. Perhaps they think they haven't got them because they have called what are in fact laws by different names. People have often played with the idea, for example, that the great overarching laws in psychology, the equivalents of conservation of energy or momentum, have gone unrecognised under the guise of 'moral principles'. What's wanted in relation to questions of free will, though, are laws less universal than (putative) ethical ones. Something equivalent to, say, Ohm's law of electricity flow would be nice. It's regarded as a good, solid law even though it could in principle be derived from more basic tenets of particle and field theory.

In fact it appears that things much like Ohm's law are known to psychology, but go under the name of attractors. As mentioned earlier in connection with the Doctor, these describe the trajectories that some complex, maybe chaotic, system can follow. In relation to neuropsychology, the complex system we are looking at is not one of ideas, as in the case of the Doctor, but of the neural activity that underpins psychology. Going up the scale, attractors also seem to play a part in social psychology, where the complex systems are not composed of individual neurons, as in neuro-psychology, but of individual people.

Attractors don't yet have the predictive power of Ohm's law because the systems in which they exist are so very complex that it is usually possible to see which ones exist in a system only by repeatedly watching how it evolves. They do have some predictive power even nowadays, though. For instance the 'strange' attractors of climatology show how you can foretell details of what can happen in a chaotic system, even though you can't forecast which particular state, out of the various possibilities, the system will end up in.

The essential point here is that the attractors embodied in a system depend, to some extent at least, on the nature of the system. A system organised in one sort of way, or containing components of a particular type, may harbour different attractors from similar systems that are otherwise organised or contain alternative components. This claim may seem opaque, but I'm really saying nothing more than that how a complex system evolves depends to some extent on its make-up. I describe it in terms of attractors in order to emphasise the law-like aspect of this fact.

Since consciousness can influence the brain, there is every reason to assume that it can also affect the attractors directing the brain's behaviour. But some of 'its' brain is actually itself. Therefore, there seems to be a sense in which consciousness may influence the laws, or at least the law-like entities, that control its own behaviour. On the other hand, maybe this argument is a bit of a verbal con trick, given that conscious experience and free will seem to have a lot more to do with information and stories than with the physical nature of the brain. It falls down, in other words, if matter and information are entirely distinct in their essential nature. You cannot light a fire with a poem, because poems and fires have different ontologies; you *can* light a fire with the paper on which the poem is written because paper and fire share an ontology. Only if no unbridgeable ontological gulf exists between matter and experienced information might consciousness influence aspects of the laws governing its own behaviour.

It could be said that speculation about the nature of matter isn't necessary. I have assumed, throughout this book, that consciousness (the expression of and contributor to the story) and some memory-related aspect of brain function (the relevant matter) can be identified with one another for all practical purposes. In other words, I have assumed in effect that the poem *is* its print on the piece of paper. Put so baldly, any such assumption seems dubious at best. But the relationship between stories and memories in brains clearly does amount to much more than that between a poem, *Paradise Lost* say, and the book in which it happens to be printed. The printing ink would not be affected if *Paradise Lost* were replaced with *Endymion*, nor would the ink used affect the content of the poem.

Stories and their neural traces, by contrast, do affect one another. This is usually supposed to be fully explainable (in principle, though not in practice) in terms of *efficient* causality. Memory formation changes neurons; changes in neurons embody memories. That is all understandable in terms of the de la Mettrie machine. I want to go a step further and suggest that the reciprocal relationship between the two surpasses the machine by also involving modifiability of *formal* causes operative in neural systems (i.e. attractors regarded as Ohm's law-like). For anything like that to be possible, stories and memories really would have to have some common ontological foundation. Neurons and memories are made of atoms and fields. Stories are made of information and meaning. The two *seem* very different. There's little to say about meaning because, as we've seen, it largely eludes scientific thinking at present. But it is not impossible that, contrary to all appearances, atoms and information will turn out to be like opposite sides of the same coin.

Physics again

At the end of the 19th century , the great thermodynamicist Lord Kelvin famously claimed that all that remained to be done in physics amounted to no more than a bookkeeping exercise. Within fifty years, Sir Arthur Eddington was making statements that to Kelvin would have sounded like the words of a mystic, while J. B. S. Haldane had said, 'the universe is not only stranger than we imagine, but stranger than we *can* imagine'. Another thirty or forty years on, and many physicists were confidently anticipating a Theory of Everything, or TOE. A TOE remains elusive and we are once more in an era of doubt.

There are various reasons for this return to the view that we may actually know very little about the true nature of matter and the Universe. One is the well-publicised incompatibility between our two most fundamental understandings, quantum theory (the physics of the smallest of scales) and general relativity (that of the largest). String theorists hope to resolve the difficulty soon, while others are more sceptical about their prospects[1]. Puzzlingly, cosmologists have discovered that the visible matter of which we and the stars are made comprises only around 5% of the mass/energy of the universe. Another 25% is 'dark matter', which interacts with our sort only via gravitation. The remaining 70% is 'dark energy', regarded as responsible for the recently discovered acceleration in the expansion of the Universe. Virtually nothing is known, so the experts currently suppose, about 95% of what exists! Even the theories about it are vague and tentative at best.

A better reason still for physicists to be modest when it comes to claiming any complete understanding of matter lies in the realisation that quantum entanglement is an all-pervasive phenomenon[2]. Entangled particles have to be regarded as single objects for some purposes, even though they may be on opposite sides of a galaxy. This may mean that neither space nor time are fundamental properties of the Universe, but derive from something else whose nature we can only guess at. Matter, it seems, is a far more wonderful and elusive phenomenon than we commonly realise.

Despite this contemporary confusion, a persistent thread has run through quantum mechanics from its beginnings. When Schrödinger first wrote down his wavefunction he thought of it as a physical wave. People very soon realised that it isn't. The preferred interpretation was to treat it as a probability wave, which when squared gives the likelihood that some particular measurement will produce a particular result. Another way of looking at the wavefunction is as a summary of all that one could ever know about a quantum object; it is a source of *information*[3]. One cannot actually get at all that information, though, as Heisenberg's uncertainty principle shows. If you measure an object's position exactly, you can discover nothing about its momentum. If you measure its spin in one direction, you can't access information about its spin in others. Nevertheless, all the information, accessible and inaccessible, is 'there', in some sense, in the wavefunction. The wavefunction is an expression of what matter *is*.

John Wheeler, famed for christening black holes in 1952, coined the phrase 'its from bits' about a decade later [4]. He meant that the particles and fields comprising matter might derive from information rather than vice versa. We usually think of particles as being fundamental, while information is some sort of derivative from their interactions. But there's a lot of evidence that we should be thinking the other way round. The further people look into the fundamental nature of matter, the more abstract and information-like it appears. A case in point is the Elitzur–Vaidman test for whether a bomb is live. For reasons that will soon become obvious, this is a thought experiment only – like the one involving Schrödinger's cat. But equivalent setups have been built and shown to behave in exactly the way predicted in the bomb test version.

The bombs in question have extremely sensitive detonators that go off if hit by a single photon. To discover which bombs are live without detonating them it is possible to set up an apparatus, involving half-silvered mirrors, which sends photons one at a time along two alternative routes. On taking one route, the photon hits the detonator; on taking the other it doesn't. The routes then rejoin. The apparatus is so constructed[5] that, when a photon arrives at the point where the two routes rejoin, it will always emerge from the junction in a particular direction if its wavefunction can take both routes. If its wavefunction cannot take both routes, the photon will only sometimes emerge in this direction (instead of always), and sometimes will pop out in another direction. This may sound unlikely, but there is overwhelming evidence that it's how quantum objects can behave. It's hard to understand, but has to be accepted. Now here's the *really* tricky part. The wavefunction will always take both routes, if it is able, just as an ocean wave arriving at a breakwater will always go round both ends of the breakwater provided the channels are open. The photon itself travels by one route or the other, but not both.

If the photon hits a live detonator, the bomb goes off and the photon is caught up in the explosion, so no photon at all emerges from the apparatus. The wavefunction can be assumed to have taken both routes up to the point the explosion occurred. If the photon goes the other way, there is no explosion and the wavefunction will take both routes provided that the bomb is a dud. If it's a dud, both channels are 'open', so to speak. The only thing that can close a channel and stop the wavefunction taking both routes is *if* the bomb *would have* exploded *had* the photon gone by the detonator route. In these circumstances, the photon can emerge from the junction in either direction. So, if the photon emerges in the direction that implies the wavefunction could only take one route, you know you've got a live bomb, but you haven't caused an explosion.

Clearly the picture of wavefunctions taking both routes and photons going either one way or the other is quite inadequate. It is a picture that predicts the results of experiments, but doesn't really make any sort of sense[6]. Nobody knows, with any certainty, what a more adequate picture would be like, but it must involve *information* somewhere. The wavefunction seems able to 'know' what would have happened if the photon had taken a route that in reality it did not take. In fact it could be entirely misleading to think of photons and wavefunctions as being either separate or real. There is information in the system (in this case about whether or not the bomb is capable of exploding), and there appear to be photons which sometimes convey that information. However, if you look at what's going on dispassionately, it must seem just as likely that the information may construct the appearance of a photon as vice versa – as Wheeler and others have proposed.

Before leaving this conundrum, it is worth noting that these 'quantum counterfactuals' not only involve information in a fundamental way, but can be viewed as having the appearance of a *story*, told by the photon: 'There I was bouncing off mirrors, and I knew only too well that, if I'd gone the other way, – boom!'. Of course the 'knowledge' attributed to the photon here is not conscious knowledge; it is more like the 'knowledgeability' of thermostats discussed in Chapter 2. Unlike the ken of thermostats, it is about what *might* happen, not about what *is* the case. The wonderful thing is that photons are essential ingredients of *all* matter (at least of the sort that we can perceive), since they are the carriers of the electromagnetic force that binds atoms together. This 'knowledgeability' or the potential for it must therefore be inherent in all matter, rather as as pan-psychists, property dualists and the like have long speculated (see Chapter 2). It's worth reiterating that, unlike many pan-psychists, I do not myself regard such knowledgeability as equivalent to consciousness.

Next, I want to mention black holes, those huge and enigmatic cosmic objects now known for sure to exist. These provide fascinating hints that the idea of information is complex and subtle; that it refers to something as robust, if not more so, as whatever underlies the notion of a material particle; finally, that it seems to be more like a process than a thing.

In the 1970s, an astonishing discovery was made[7]. Black holes have a well-defined entropy, which is a measure of the amount of disorder in a system. This was surprising for three reasons. First, people had previously thought that you could know nothing whatsoever about a black hole except for its overall mass, electric charge and spin (angular momentum). Second, the equation describing the entropy shows that what matters is not the volume of the hole, as one might expect, but the area of its event horizon. Third, the actual figure for the entropy was vast, relating to areas on a scale many billions of times tinier than the scale on which sub-atomic particles (of the types we know about) exist. The finding was much criticised at first, but has stood the test of time and is now generally accepted.

What has all this to do with information? Well, one concept of information is exactly equivalent, for all practical purposes, to that of entropy. This is algorithmic information, a variant of Shannon information (our bits and bytes). People often think that there is only one notion of information, but, apart from these two types, there is also Fisher information, a quantified measure of our degree of ignorance of a phenomenon, and more general concepts such as Gregory Bateson's, which has cropped up previously. The idea behind algorithmic information is that the more complex a system, the larger the computer program needed to describe it. If a system is completely random, the program needed to describe it must contain at least as much information as the system itself. So, in effect, black holes must also be regarded as having an enormous (algorithmic) information content along with their huge entropy. There's a caveat here, since the same does not necessarily apply to other concepts of information. It used to be supposed that Shannon information gets destroyed in black holes. Nowadays Stephen Hawking concedes that it isn't destroyed. Fisher information, on the other hand, must indeed be lost forever when something falls into a hole. It is a measure of how much we know about something, and it's absolutely impossible to know anything about objects once they have passed the hole's event horizon.

Whatever the case with other types, if algorithmic information, or indeed entropy, were somehow directly equivalent to physical objects, then everything falling into the hole – stars, meteorites, photons – would carry some actual or potential entropy/information with it. The quantity of objects packed into a hole is certainly proportional to its volume, so why is algorithmic information proportional to area? Surfaces separate different zones, in this case the inside of the black hole from the rest of the Universe. That the surface defines an amount of information contained in the hole may be telling us that relationship is part of its fundamental nature, which is fully consistent with Bateson's beautiful and profound definition ('a difference that makes a difference'). It is not a thing but is story-like. And it seems to be preserved in some sense even when objects are lost or destroyed within a hole. So one may wonder (as loop quantum gravity theorists have wondered) whether something information-like may be fundamental, which generates a real, material world rather as an optical hologram can generate the *illusion* of a real world.

There are many unknowns, and this section is about the sort of interpretation that may be put on fundamental physics one day, so it would be rash to draw any very firm conclusions. All the same, it does appear more likely than not that there is no absolute ontological separation between matter and information, while information itself may be inherently story-like. The next step – the final leg of the rickety tripod – is to see whether there is any evidence at all for my line of speculation. Are there any indications that memories can convert themselves into law-like entities – attractors – capable of influencing the future course and content of consciousness? It's worth reminding you that I'm not just looking for an 'efficient cause' capability, since clearly memories can influence the future of their brains via such causes. We've already established that in previous chapters. I'm looking for evidence that conscious memories can influence the very laws by which they operate.

A captain of the ship with this capability would no longer be a sophisticated autopilot, he would be changing the physics governing his own behaviour – possibly in a manner that accorded with his own intentions at times. He would be at least as different from any of our existing machines as my laptop is from a thermostat. Unlike a thermostat, a laptop can be made to program itself (to a limited extent), and so can our brains via a range of mechanisms dependent on efficient causes. They are much more capable in this way than are laptops. But the captain I am envisaging would be able to alter the physical characteristics of his own hardware – something that no computer can achieve. De la Mettrie, confronted with an entity like this, would have had a lot of difficulty in regarding him/it as a machine.

Surely we already know that brains can alter their hardware? After all, their own activities contribute to switching on and off the genes that influence the structure of their constituent neurons. People can choose to swallow intoxicants or undertake special training, either of which may greatly affect how their neurons behave and can even modify the anatomy of their brains. True enough, but capabilities like these are still within the realm of efficient causes. De la Mettrie's metaphor would need a lot of stretching, but could still just about be made to apply to such abilities by a sufficiently diehard reductionist. Genetic switches and the like are getting close, but do not yet introduce a possibility of freedom in any ultimate sense. For that, an influence of consciousness on law-like determinants of its behaviour is needed. If this exists, what sort of evidence for it might we look for?

It has to be subjective evidence because we are dealing with consciousness here, which is inherently subjective. And my problem is not that there is a shortage of potential evidence. Rather, there is an embarrassing excess, but it is almost all garbled, unreliable and hard to interpret. Evidence may be present in 'exceptional human experiences', which range from alien abduction through past life 'memories' to clairvoyance and the like. In fact there are said to be more than eighty subtypes of exceptional experience, classifiable into five main groups. I'm going to take a look at what just two of these subtypes can tell us. They are among the best and most reliably documented of any in this confusing field.

Subjective experience

Doctors have always considered that reports of subjective experience, however bizarre, yield valid and useful evidence. Non-medical scientists are far more hesitant. For most psychologists and neuroscientists, the use of subjective evidence was heretical while Behaviourism reigned in the mid-20th century. The pendulum has not yet recovered fully from this extreme swing. As a medical doctor myself, I regard any moratorium on subjective evidence as ludicrous. But there are excellent reasons to be extremely cautious as to how it is used, and even better reason to be wary when one comes to deciding what meaning should be attributed to it. With these caveats in mind, I'll describe the two types of exceptional experience that I've selected and see what conclusions can be drawn. First, death:

Near-death experiences

These have received a lot of attention since the philosopher and psychiatrist Raymond Moody published his hit book *Life After Life* describing them in 1975. His interest was sparked when some of his philosophy students described NDEs to him. He continued to collect accounts after switching to medicine, and had garnered around 150 by 1974. Large numbers of people have continued to report these experiences. There is an estimate that as many as 10% of Americans have had an NDE or 'significant' out-of-body experience. This figure seems rather high, unless 'significant' is interpreted generously. Only around 10–20% of people needing resuscitation after heart attacks report even partial NDEs, let alone the full-blown experience.

The experience usually begins, people claim, with feeling calm, peacefulness and sometimes joy, together with loss of any pain. Typically, they then describe getting detached from their body and floating usually upwards towards the ceiling[8]. Later most claim to have passed through a tunnel towards a light. At the end of the tunnel, they may encounter living or dead relatives, or other figures, sometimes in the setting of beautiful gardens or landscapes. Some undergo a 'life review' in which they re-experience, with great intensity and within a seemingly short period of time, all the good and bad things they have done. A few of these 'life reviewers' have claimed also to experience events from the point of view of people they had harmed. Then they become aware that they must return to their bodies and find themselves 'waking up' to whatever, generally distressing, situation they had left. The calm and apparent clarity of the NDE is lost, and confusion and pain return. Not every NDEer undergoes the complete range of experiences. A small proportion find themselves in hellish surroundings. People from all cultures have reported NDEs, but the content of some of the experiences seems to be culture-related. For instance, a Zambian woman is said to have experienced entering a calabash rather than a tunnel. Christians are likely to meet Jesus, whereas Hindus may encounter Krishna.

There have been lots of attempts to explain the phenomenon in causal, scientific terms. Temporal lobe epilepsy, the neurotransmitter glutamine, endorphins and dimethyltryptamine (DMT)[9] have all been proposed as possible causes at various times. Susan Blackmore has developed an especially elaborate model, involving anoxia and the brain's dying attempts to restore a sense of reality in the best way it can. As Peter and Elizabeth Fenwick (a husband-and-wife team with a vast experience of neuropsychiatry between them) argue, though, it is hard to reconcile these explanations with the enhancement of consciousness and clarity of memory that many NDEers claim. When people are dying, or having a fit, or anoxic, the usual result is that they get confused and forgetful, not extra lucid with super-normal recall of their past. DMT or some similar hallucinogen might on its own induce heightened consciousness, but it is not easy to believe that it could do so when associated with events of the type that often cause NDEs. Blackmore has pointed out that not all NDEs accompany events that are in reality life-threatening. Similarly, other 'out of body' experiences, while often apparently due to severe psychological or physical stress, sometimes just happen for no obvious reason. But these considerations do not help to explain how the ones that are associated with life-threatening illness could achieve such apparently remarkable clarity.

Many NDEers themselves go along with the picture of their souls drifting ceilingward from their bodies, then getting sucked up by some sort of cosmic vacuum cleaner and deposited in the suburbs of paradise. That's what they experience, so why should they doubt it? Moreover the experience is transformative for some. There are many reports of people losing all fear of death, becoming less materialistic and more loving, following an NDE. Judging by these after-effects, the experiences must have a very unusual intensity and ability to lodge in memory, just as NDEers say. The picture of souls drifting away from bodies won't do, though. It is rather reminiscent of the picture of photons and wavefunctions going their separate ways through an apparatus. It accounts for the phenomena all right, but doesn't fit well with nearly everything else that we know. Further, it embodies all the dilemmas of Cartesian dualism in a particularly acute form. How could immaterial souls possibly influence a material body without breaking all sorts of laws, especially that of conservation of energy? Why, in the case of NDEs, should an immaterial soul be able to impress such remarkably clear and vivid memories on a damaged body that has a hard enough time hanging on to even confused memories of its own?

Clearly NDE experiences, like all others, must be associated with – if not necessarily 'caused by' – distinctive neural states. These neural states will themselves have efficient causes, which may include any or all of those already suggested in scientific explanations of the phenomenon. What is missing is an account of, first, why NDEs have the content that they do and, second, why intensity and clarity are found where confusion and clouding would be expected. The content is actually a sort of story, made up from a range of elements, some personal and some relating to family and culture. There are ideas about the nature of the self or soul, expressions of ethical and religious concepts, and memories both of personal events and significant people in one's life. What seems to happen in a well-developed NDE is that people experience a sort of biopic novel or autobiography about themselves and their histories, but one that is 'played' extraordinarily quickly, often in circumstances not at all conducive to watching anything. 'Anoxia', or whatever neural efficient cause might be proposed, is clearly inadequate to account for the development of the very special and precisely organised neural states that must accompany NDEs.

The most straightforward explanation for this specialness is that the stories themselves can act like attractors in a chaotic system, and formally cause the appropriate neural states to crystallise out. At times of crisis, this argument goes, the stories most deeply ingrained in a person are likely to provide the most powerful attraction for all the neural chaos going on and so, perhaps, may be the ones experienced.

Ayahuasca

The experience of ayahuasca intoxication has been beautifully documented by Benny Shanon[10], a cognitive psychologist and professor at the Hebrew University of Jerusalem. Shanon took the drug on more than 67 occasions and gathered evidence from other people, both anecdotal and from structured questionnaires, relating to another 2500 sessions. It is a substance not to be taken lightly. Jeremy Narby, an anthropologist with personal experience of it remarked, 'To my mind, a truly hallucinatory session is more like a controlled nightmare than a form of recreation and demands know-how, discipline, and courage'[11].

Ayahuasca is an Amazonian brew, made in quite a complicated process from two different varieties of plant, traditionally used by tribal shamans. It is now adopted by a number of religious sects, which incorporate Catholic ideas along with shamanistic and other traditions. It can produce extremely intense and fantastical experiences, which have quite a bit in common with some near-death experiences. This may be more than a coincidence, since one of the ingredients in the brew is DMT, a chemical which can also be produced by the brain from its own resources (see note 9) and has been proposed as a possible cause of NDEs. Ayahuasca contains other psychoactive substances: the harma alkaloids. These are mildly hallucinogenic, and block the enzymes which would normally destroy any ingested DMT before it could reach the brain. While DMT certainly looks like the most important ingredient according to Western ideas, some shamans regard the plant containing the DMT as the adjuvant and constituents of the brew derived from the other plant as the essential ingredients.

The brew tastes disgusting. It quite commonly induces a range of physical sensations so horrible that drinkers feel as if they are about to die, though apparently it is actually quite safe if taken under supervision. Hallucinations induced by the drug sometimes start off with migraine-like geometrical shapes and lights. There is no headache, though sickness and stomach upset are common. Sooner or later drinkers enter a magical world of forest scenes, gardens and/or fantastic cities. Says Shanon:

> ... it is as if out there life goes on and the intoxicated person is presented for a moment with the opportunity to witness a scene of other times, and other places. In glimpses this fragment of time is very short, in full-fledged scenes it may be long. The historical scenes may depict major historical events or episodes of daily life. The most common instances of the former are wars, coronations and royal pageants, and episodes in the lives of famous historical figures. Images of

> the latter kind that stand out in my mind are a market scene in Mediaeval Europe and a very colourful street scene in China...

He goes on to report that people often feel that they are experiencing something 'real', even when they transform into an animal, bird, or plant:

> ... Indians I talked to said... that Ayahuasca enables them to travel and see foreign places. A lower-class, simple resident of the far west of Brazil told me that he had visions in which he found himself in Russia and Japan. In real life, this man had never left the Amazon region. Further, he assured me that he had never seen pictures of these places.

Apparently the types of vision seen tend to alter as time goes on. Personal, biographical material is especially common in drinkers' first sessions with ayahuasca. The more experienced they are, and the more powerful their visions, the less idiosyncratic and personal the contents of the visions.

Shanon often uses the phrase 'the school of Ayahuasca', apropos his feeling that the drug tends to induce whatever experiences may be best suited to promoting the spiritual health of drinkers. Philosophical 'insights' can also occur. Shanon experienced one such, concerning choice:

> In front of me I saw the space of all possibilities, that is, all states of affairs that can possibly happen. They were lying there in front of me like objects in physical space. Choosing, I realised, is tantamount to the taking of a particular path in this space. It does not, however, consist in the generation of an intrinsically new state of affairs. All possibilities are already there, I saw, but one has the option of choosing different paths among them...

This seems to be a visualisation of much the same concept as that Hodgson described in his 'superlife universe' (see Chapter 4).

These experiences have all the characteristics of fairy tales. The obvious explanation for users feeling they access something real and outside of themselves is that they are in a temporarily psychotic state; loss of insight into what is hallucinatory and what is not often occurs in psychosis[12]. The experiences of the Indian who claimed to have visited Russia, for instance, may have been dream-like fantasies. If their content did resemble the real Russia in any respects, the correspondence may have been based on unrecallable memories of information that he had had access to at some time. It's virtually impossible to prove that someone has never had access to information of this sort.

More interesting is some suggestive, though far from conclusive, evidence that these strange experiences can possess a supra-personal, law-like existence. It is a bit like the slight discrepancy in the precession of Mercury, which told Victorian astronomers that there could be something wrong with their Newtonian concept of gravity. The discrepancy was so undramatic, compared to all the other wonders of astronomy, that they were not greatly disturbed by it[13]. Similarly, there is a discrepancy in the phenomenology of the Ayahauasca experience that is not dramatic, but does seem to be pointing to something very odd indeed. It is to do with seeing pumas – the big cats, that is.

Shanon again:

> In the literature, apart from serpents, the most common type of animal associated with Ayahuasca visions are felines, notably jaguars and pumas. This is also the case with respect to the data that I have collected, both mine and those of my informants.... In my structured interviews I have asked my informants whether they have seen 'jaguars'. Again and again, the response I have received was 'Yes, but more frequently I have seen pumas, black pumas'.

The pumas can appear, says Shanon, whatever the race or experience of the drinker and regardless of where the session occurs. But why should European, first-time drinkers see South American animals, especially if they were not even in South America at the time? If the animals were there in their role as predators, surely a European might be expected to see wolves or bears most frequently. If they were there as representatives of big cats, then lions and tigers are far more familiar to most of us and would be expected to come to mind first. Maybe, through some physiological quirk perhaps, they have to be big cats who are either black and white (jaguars) or all black (black pumas). Well, there are no all-black big cats outside South America, but there are much more familiar black and white ones than jaguars, i.e. leopards. Shanon does report occasional visions of both tigers and cheetahs. But the curious fact remains that people are more likely to see black pumas, even when common sense would tell us that they ought to be seeing something else. These observations seem so counter-intuitive that it is surely appropriate to suppose they might be hinting at something very surprising and profound about the origins of the content of experience.

So people who are extremely unlikely to have any deeply ingrained memories of black pumas see them under the influence of ayahuasca, in preference to equivalent animals that are far more likely to have a place in their memories. It is hard to believe that efficient causes alone could produce a result like this. Surely the stories most prominent in anyone's mind, and most likely to emerge under the guidance of attractors in the course of the relative chaos induced by intoxicants, should be those belonging to one's personal history, social circumstances and culture. Unless, that is, the behaviour of attractors is *not* fully determined by efficient causes, or at least not by such causes as they are currently understood.

Y

The NDE experience hints that consciousness may influence, no doubt via its links with memory, the attractors that affect its future behaviour. Although a suggestion like this is getting a bit ahead of anything that can confidently be concluded on the basis of contemporary neuroscience, it is not all that surprising. The attractors in question could be emergent properties of interactions among the relevant efficient causes – the neural bases of all the memories that have gone into the make-up of some particular person – rather as the 'laws' of hydraulics arise from interactions of water molecules. On the other hand, the possible implications of the black pumas are extremely surprising. Maybe they do provide some sort of valid indication as to how we should be thinking in the future. They seem to suggest that attractors, of the sort suggested by the NDE experience, may sometimes be supra-personal and thus more law-like than we can currently countenance.

Actually, such a suggestion is far from new. Jung's idea of archetypes, for instance, was very close to the modern concept of an attractor. Jungian archetypes are notions shared by the whole of humanity, such as the Mother or the Sky Father. To be fully accurate, I should point out that he regarded the tendency to develop such a notion as the archetype itself, and the actual notion appearing in someone's experience as an 'archetypal representation'. After years of speculation, mostly in the 1940s and early 1950s, the great physicist Wolfgang Pauli concluded that he couldn't explain his friend's archetypes in terms of quantum theory, as Jung had hoped might be possible. He hedged quite a bit because Jung was the older man and had been Pauli's therapist earlier on, but his conclusion was firm. Fifty years on it still holds. On the other hand, there do seem to be grounds now for thinking the situation could change when quantum theory itself is completed or superseded.

Another problem Jung never solved was the question of whether or not, to use modern terminology, his archetypes were emergent properties of the underlying human biology, or whether they were transcendental objects existing in Platonic heaven, like the 'perfect' circles or triangles that some mathematicians believe to exist in a deeper reality underpinning our phenomenal world. He oscillated between these two views at different stages of his life, sometimes favouring one, sometimes the other, and was liable to get angry if reminded of his former fancy. But he may have been troubling himself unnecessarily: the dichotomy could well be false. Human biology and transcendental objects may be different sides of the same coin. To understand Jungian archetypes, and the attractors in memory, we may need to study both their biology and their transcendental, quasi-mathematical aspects.

If attractors are indeed law-like then determinism, at least as far as brains are concerned, has to be regarded as a larger and more fluid concept than current orthodoxy allows. Minds may be able to influence the formal causes, as well as the efficient ones, that control their behaviour. And formal causes are supra-personal things. If meaning is more a property of the network of story lines than of the individual, as was concluded in Chapter 12, it too is supra-personal in a sense. Could the pumas be pointing to a route that may one day allow scientists to get a firm grip on the real basis of the origins of meaning? Many people wonder whether matter, information and mind are more closely related than is generally supposed at present[14]. Developments in fundamental physics should one day allow us to do more than wonder. Physics, however, must work from the material end of any equation. It would be good to follow up hints allowing us to approach the puzzle from the mental end as well – and so help physicists with their explorations of the foundations of reality.

Epilogue

ON BEING A STORY

One of the main reasons for the confusion surrounding free will is the nature of the 'I' that makes choices. Naturally people tend to think of 'I' as the conscious self-awareness at the moment of choice, or even as the conscious output of the 'ownership of action module' that we met in Chapter 5. In fact what does the choosing is a temporally extended 'I' that we never get to experience all at once. It is an agglomeration of all the stories accumulated in one's memory. Not all of them contribute to each choice, but a proportion do. Memories from forty years back may influence the outcome of even quite trivial choices just as much as memories of current situations. These memories will nearly all have been either conscious or selected by consciousness at some time, but most won't be consciously recalled at choosing time. For example, my deciding whether to have a cup of tea instead of coffee just now may depend on what I consciously recall of my feelings of thirst, tiredness and so forth over the past couple of minutes, but will also depend to a considerable extent on my lifetime's experience of these drinks. Moreover, the enactment of choice always involves all sorts of unconscious neural activity, some of which never gets to be conscious. The detail of controlling eye muscles, for example, which is integral to many types of conscious choice, is always unconscious. You may consciously decide to look left but you never know how your opposing sets of eye muscles work to obey this command – which is just as well. If you were aware of the detail, there would be little room in consciousness for anything else, as it can hold relatively little information at any given moment.

To see who it is that has free will involves adopting a quite different perspective from the one habitual to most of us. We must view ourselves as a collection of stories embodied in memory, not as de la Mettrie machines. Most of the stories don't even derive from our own personal histories. They originate in our societies. We do have some choice about which ones we incorporate in ourselves, though the majority get into us whether we like it or not via education in the broadest sense. Moreover they don't necessarily behave as we might like once they do get in, as the Dance and the Saint illustrate. Perhaps even the best of saints have wished they could have been nicer to their families.

An extraordinary example of the all-pervasive influence of stories on our lives can be found in the recent empowerment of western women. The example also shows nicely the rather long timescales that can be involved. Despite all the fleeting swings of fashion that come and go, the really important story-enabled changes in society may occur over generations and centuries rather than years or even decades.

Back in classical times you pretty much had to be either a Goddess like Athene or closely associated with a male hero in order to attract much general notice. The poetess Sappho was very much the exception proving the rule in being both human and her own woman. Things only got worse at the end of the period. The picture of the last female academic at the great Museum (actually more of a university) of Alexandria being hounded to death by a mob urged on by Bishop Cyril is probably fairly accurate. Her name was Hypatia, and we hear of no more learned women in public positions for a long time subsequently. Cyril, incidentally, later became St Cyril, which says a lot about the values held by the powers-that-were at that time. Clearly they did not include an appreciation of the worth of people like Hypatia. For more than a thousand years subsequently in the West, nearly all the 'official' stories were told by men, though no doubt there was an undercurrent of persistent, but probably very localised and relatively fragmented, female lore. The only women who could attain wide prominence in that period were those who happened to fit some primarily male fantasy or story, like the Wise and Holy Virgin (Hildegard of Bingen), the Object of Courtly Desire (Eleanor of Aquitaine and many others) or the Virgin Queen (Elizabeth I).

Then things changed. Maybe imaginations had been fired by figures like Chaucer's Wife of Bath and Shakespeare's Portia. For whatever reason, towards the end of the 17th century women like Aphra Behn began to tell their own stories in their own way and gained access to the public arena so that their stories could spread abroad. The process was slow and hesitant at first, with several false dawns. But it slowly gathered momentum through the 18th and 19th centuries with Mary Wollstonecraft and her daughter Mary Shelley, for example. Then there was the suffragette movement and the two world wars, which proved to women, and some men at least, that women were at least as competent as men in all sorts of public as well as private roles. Now both women and their stories have achieved a modicum of parity with men and theirs in many areas, to the great enrichment of society and all our lives. Maybe there's still a way to go in some fields, but nevertheless women now have opportunities and lead lives that would have been very hard to imagine even a hundred years ago – a change primarily due to interactions between them and the stories available to them, though no doubt facilitated by mundane things like better contraception and the invention of washing machines.

And yes, the overall picture of society that I've developed here owes a great deal to the visionary account provided by a female writer, Doris Lessing. She herself probably derived some of her thinking from male sociologists like Emile Durkheim, but rather more, one suspects, from female anthropologists like Margaret Mead or Ruth Benedict. Lessing certainly fits one male story, that of the Wise Woman, but she is fiercely independent. Her *Canopus in Argos* series of books give vivid allegories of exactly the sort of concept of the social network that I have tried to convey in relatively pedestrian and scientific terms here. And the books have such wonderful embellishments, too. My personal favourite is the 'hospital for rhetorical diseases' featured in *The Sentimental Agents in the Volyen Empire*. Those suffering from an excessive tendency to utter or believe rhetoric are confined to a luxurious room with magnificent views of a darkling forest, backed by looming mountains lit with lurid sunsets in one direction and huge waves crashing against towering cliffs in another, while Beethoven and Wagner are played at full volume. Their natural reaction to these excesses cures some of them. If only, one cannot help wishing, our own dear politicians could be treated so.

Y

Margaret Thatcher, no mean rhetorician herself and wrong about many things, was utterly wrong when she said, 'there is no such thing as society'. As far as our conscious experience goes, which after all is the only reality we have, there is little else. We would be almost entirely empty vessels if the network did not exist. In the absence of society we should be speechless and nearly mindless wild children. Henry Ford was equally wrong with his 'history is bunk'[1]. It turns out that all but the most basic aspects of our lives embody long histories[2], many of them with origins in Roman times or earlier. And it's not just a matter of recalling tales or ways of thinking in some sort of abstract sense. Rather, the stories become part of us. All of us sense this in the importance we attach to family histories, national culture, the Western or Eastern heritage, and so forth. However, it is a sense which has become quite marginalised in recent times, partly because the de la Mettrie machine has been centre stage for so long.

It has been said that poets are the 'unacknowledged legislators of the world'. The view I'm proposing suggests that this must be true of all storytellers; not only novelists and film makers, but artists in general, teachers, makers of theories, and so forth. We are all storytellers to an extent, but some people have many more links in the social network than others; these people are likely to be especially influential. Stories create our personhood and our worlds.

This fact carries a warning for us. There is a magnificent range of stories to be found in educational material, literature and film that is supposedly available to all. However, for every purveyor of what is worthwhile, be it a good schoolteacher, Jane Austen, Spielberg or Tolkien, there are hundreds of writers of impoverished tales in the tabloid press, purveyors of TV banalities, trashy movies, sick computer games and the like. In the short term these debased stories may do little harm, but the cumulative effect over generations will not be so harmless. At least, it will not be harmless if the picture of personhood and choice that I have sketched here is even half right. People absorbing impoverished and plain nasty stories will increasingly as time passes tend to embody these qualities.

Of course all human life has to be represented in stories, so *some* video nasties are probably necessary, just as there are arguments for conserving mosquitoes as well as elephants. The traditional literature on which most societies have been nurtured, the Bible for instance, contains its fair share of gruesome and apparently unedifying tales. But the share does have to be fair, not excessive. Luckily there are wonderful artists around to put magic, complexity and virtue back into our lives, while science generates an ever-increasing fund of *true* stories. It is a pity, though, that it is not so easy nowadays for what is beautiful, true or subtle to make itself visible through all the spam.

Shakespeare, that most perceptive of great storytellers, got it right as usual:

> ... 'tis in ourselves that we are thus or thus. Our bodies are gardens, to the which our wills are gardeners; so that if we will plant nettles or sow lettuce, set hyssop and weed up thyme, supply it with one gender of herbs or distract it with many, either to have it sterile with idleness or manured with industry—why, the power and corrigible authority of

> this lies in our wills. (Speech by Iago from *Othello*, Act I, Scene 3)

But of course we can only plant what comes to hand – and if nettles are far more readily available than useful herbs...

This is not a call for censorship in general, though it is a strong argument for making every effort to limit children's exposure to impoverished, depraved or violent tales, especially those that they themselves are made to enact in the course of certain computer games. The sad history of eugenics showed that trying to censor 'undesirable' genes was a disaster, and the same might well apply to even the most well-meaning of blanket efforts to control unpleasant stories. Interventions of this sort are only too liable to achieve the opposite effect to that intended.

Better would be the reverse of censorship. Even greater efforts than those already being made are needed to nurture the creation and promulgation of the widest variety of stories possible, not just sex and violence. As the popularity of Bollywood epics, TV soaps and fantasy blockbusters shows, people have an enormous appetite for well-told tales, whatever their content. Moreover, most children, as well as some adults, have an equal appetite for education unless deterred by bad teaching. We all, especially when children, need ready access to the good stuff as well as to the 'chicken nuggets' of narrative. Only then is it possible to make the choices that can lead to a healthy future.

NOTES

Chapter 1

1. Actually, many of them, including Isaac Newton, are more correctly referred to as 'deists' than as 'theists'. The former envisage a God who set the apparatus of the universe in motion, and who also perhaps functions as a 'God of the gaps'. 'Theists', on the other hand, believe in an immanent God. An excellent account of these issues can be found in Alexander's book (see References).

2. David Chalmers is a philosopher who worked for a time in Arizona, but is now back in Australia. He has published widely on consciousness-related issues, and is probably best known for his book *The Conscious Mind*, OUP, 1996. He has also been one of the chief organisers of the Tucson conferences. These are large gatherings of people interested in consciousness studies, held every two years.

3. It has been argued that people can legitimately be regarded as responsible for their actions, even if their behaviour is in fact fully determined by physical law and by chance. See, for example, the paper by Freeman listed in the references. The picture to be developed in this book is rather different. While agreeing that free choice may be grounded in law and chance, I will show that the notion of free, conscious choice is nevertheless valid.

Chapter 2

1. Karl Popper wrote a book with the Nobel-prize-winning neurophysiologist John Eccles, in which they argued that elementary, non-material entities called 'psychons' influence quantum-level events in synapses.

2. Idealism is the view that the whole of reality is fundamentally 'mental'. The 18th century philosopher Bishop George Berkeley is the best-known exponent of this view, holding that an oak tree for instance exists only because it is held in the mind of God. He is also

well known for his very cogent, but critical, description of Newton's infinitesimals as 'the ghosts of departed quantities'. Idealists, therefore, agree with materialists that there is only one substance, but have their own opinion about its fundamental nature. As Galen Strawson has pointed out, idealism is best regarded as no more than the opposite side of the coin of materialism.

3. Francisco Varela, who died in 2001, was a particularly effective advocate of the idea that mind must always be considered in the context of its embodiment.

4. A number of people have suggested that the informational aspects of quantum theory strongly suggest that pan-psychism has to be taken seriously (e.g. Seager). David Skrbina has provided a nice, brief account of the history of pan-psychism in Western thought. Herms Romijn, a Dutch neurophysiologist, may have been the first to describe the proposal that a pan-psychist type of awareness is embodied in virtual photons, in an unpublished manuscript written around 1995.

5. This is the famous 40 Hz coherent activity which appears in separate brain areas when they can be inferred to be working on a single percept. This observation has held up fairly well, though the range of frequencies involved is far greater than was at first supposed (probably from 30 Hz to maybe as much as 80 Hz). Moreover, neural synchrony or coherence does not always seem to occur in relation to what might be assumed to be single percepts, even though it usually does (for contrary findings see the papers by Thiele and Stoner and by Lamme and Spekreijse in the references). There is an interesting paper by Cossart and colleagues, showing how the coherent activity may arise.

6. The most numerous of the cells that support neurons are called astroglia. Waves of calcium ions within them, which probably spread from cell to cell via gap junctions, are thought to be capable of enhancing synaptic transmission. Nitric oxide is probably important to the control of blood flow through the brain, but it may also help to modulate synaptic function and thus be directly important in how the brain deals with information.

7. One of the first of these 'others' was Ian Marshall in 1989, who proposed that Bose–Einstein condensation might solve the binding problem. A Bose–Einstein condensate is a unified quantum field of

a similar type to the one responsible for (low temperature) superconductivity.

8. This theory was developed by Stuart Hameroff, an anaesthetist, working with Sir Roger Penrose. It is referred to as the 'OrchOR', or orchestrated objective reduction, hypothesis because it incorporates Penrose's ideas about gravitational responsibility for the collapse of quantum wavefunctions. A clear, brief account of their ideas can be found in their paper cited in the references.
9. There are quite a lot of suggestions of this type. One with a better pedigree than most is based on the field theory developed by the Japanese physicist Hiroomi Umezawa, described in Jibu and Yasue's book. A similar line of enquiry is still under active consideration, especially by Giuseppe Vitiello.
10. A particularly striking demonstration of the fact that consciousness must be related to (very short-term) memory is provided by the visual system. Higher visual processing is divided into two main streams; a dorsal 'where is it?' stream and a ventral 'what is it?' one. Part of the dorsal stream is a sort of rapid reaction system, which enables one to grasp objects quickly for example. Unlike the ventral stream and other aspects of the dorsal stream, which take longer to process information, thus giving memory time to come into play, the rapid reaction system is unconscious. As a consequence, bizarre phenomena can occur like consciously perceiving an object to be of some illusory size, but nevertheless unconsciously adjusting one's grasp correctly in relation to its actual size.
11. People have pointed out that the reciprocal of this statement is also true. There's a lot of evidence that recurrent activity on all sorts of scales is at the basis of brain functioning. This fact, it has been said, is likely to prove to be at the basis of short-term memory.

Chapter 3

1. Korsakov states are relatively rare these days. They are usually due to vitamin B deficiencies associated with alcoholism. They were not uncommon when some alcoholics tended to get all their calories from cheap spirits and ate practically nothing. Beer drinkers were relatively safe, as beer contains some B vitamins. Nowadays better nutrition and earlier treatment have practically abolished the very severe memory loss that used to be seen, though milder forms still

occur. The best descriptions of the 'full blown' form are therefore to be found in older (1960s vintage) textbooks of psychiatry.

2. It's worth noting, though, that hippocampal functions are still regarded as secondary by some. Watt and Pincus, for instance, note that there are current theories which give a central role to the cells that cover the surface of the thalamus. They comment that there are proposals this area 'functions as a net on which potential working memories and conscious content compete, proposing that material makes it into working memory by virtue of potentially widespread "alliances" established on its surface...'.

3. Quite a lot of interest has more recently been taken in the analogous capacity in animals. It has been inferred that most birds and mammals can hold at least three items in immediate memory, and some can do rather better than this.

4. It seems that visual working memory involves a rather different brain area from most types, namely the posterior parietal cortex. Maybe it is this difference that accounts for the smaller capacity of visual memory for separate items. Interestingly, the more visual items a person is asked to remember, the harder their parietal cortex has to work; apparently it soon 'runs out of steam'.

5. Nicolis and Tsuda put it like this in their paper: '... the "human channel", which is so narrow and so noisy (of the order of just a few bits per second or a few bits per category) possesses the ability of squeezing... practically an unlimited number of bits per symbol – thereby giving rise to a phenomenal memory'.

6. The overall probability is the product of a whole lot of separate probabilities. An action potential may or may not release any neurotransmitter. If it does release any, there is a question of how much. The neurotransmitter may or may not cause the next cell to fire. There is also a question of how much change in the cell's membrane potential a given amount of neurotransmitter may cause, which is also relevant to any future firing probability.

7. Although known to occur, it was surprisingly difficult to establish for certain that long-term potentiation had any relationship to memory and learning in practice. This has now been achieved, though not until 1997 as reported by Malenka and Nicoll. LTP involving NMDA receptors may be the most important type in the hippocampus, but other types of LTP, not involving these particular

receptors, also exist. The opposite phenomenon, long-term depression, also occurs, though its role is even less well understood than that of LTP. There are good reasons for thinking that anatomical changes in synaptic spines, dendrites and whole neurons are also important to long-term memory; perhaps more important than synaptic changes, whose main role may lie in triggering more permanent structural alterations.

8. An example of the sort of speculation that can be indulged is the following. Karl Pribram has for thirty years been advocating the idea that many brain functions, especially memory, are based on holographic principles. CaMKII provides an ideal recording medium for interference fringes between waves of varying calcium concentration. Recording interference fringes is at the basis of any holography. Moreover, it is possible to imagine that CaMKII-based holograms might be 'self-playing', so to speak, as a consequence of the protein's effects on synaptic efficiency. In other words, they might not need the equivalent of illumination by an external light source that is needed to see an optical hologram. A fuller account can be found in Nunn, C. (2003) A Nagelian neurology of consciousness? *Science and Consciousness Review*, http://psych.pomona.edu/. I hasten to add that this speculation cannot be blamed on Pribram, who has other ideas about the likely role of holograms in mentality.

Chapter 4

1. This appeared in a collection of essays that Hofstadter wrote, many of which originally appeared in *Scientific American*, after achieving fame with *Gödel, Escher, Bach*. Hofstadter later became increasingly interested in wordplay. His most recent book, *Le Ton Beau de Marot*, is a 600 page exploration of the subtleties involved in translating a (quite short) French poem.

2. The part of the frontal lobe chiefly involved is called the dorso-lateral pre-frontal cortex. Together with its feedback loops to other cortical areas and to sub-cortical centres, it seems to play a principal role in the 'voluntary' choice of action.

3. William James, brother of Henry the novelist, was a psychologist and philosopher who made pioneering observations on consciousness, among other subjects. He is best known for two books, *The Principles of Psychology* and *The Varieties of Religious Experience*. His

influence on psychology, which was great in the 19th century, rapidly waned in the 20th due to the swing of intellectual fashion towards Behaviourism. After nearly a century of neglect, we are only now beginning once more to appreciate how remarkable and prescient a writer he was.

4. This was discovered by Grey Walter working in Bristol, though he rarely gets a mention in recent literature on the subject, the credit often going to Kornhuber and Deecke instead. Grey Walter called the electrical change an 'expectancy wave', though it has since been renamed 'contingent negative variation' (CNV). The method he used was to tell people that they were to press a button when a buzzer sounded. The buzzer was preceded by a warning light. He recorded their EEGs (brain electrical activity picked up by electrodes on the scalp) from the time the light came on till they pressed the button, and then obtained the average EEG from repeated trials. In this way, random EEG activity cancelled out and he was able to see the task-related electrical change, i.e. the 'expectancy wave'. He used this very restricted methodology because the technology of the time did not allow him to average more than a few seconds worth of EEG recording per trial. Libet, working nearly twenty years later, was able to average records over longer periods.

5. I am referring here to the Princeton PEAR project conducted by Jahn and Dunne, in which people have tried to affect the output of random number generators with their conscious wishes. Over a huge number of trials, statistically very significant effects have been found. However, the chance of consciousness affecting any individual trial is minute, so tiny that it is hardly a promising basis on which to ground any notion of free will. See, for example, the book by Dean Radin. Although the Princeton researchers have been able to replicate their findings, an attempt to do so by German researchers failed. This is a common problem in 'psi' research. It is almost as if anomalous findings invoke some sort of corrective mechanism which returns everything to a random condition overall. This makes it even less likely, of course, that 'psi' effects could be at the basis of free will.

6. This was the interpretation that gave rise to the oft-told tale of Schrödinger's cat. The poor beast existed in a 'superposition' of states in which it was both alive and dead until somebody looked inside its box and 'collapsed its wavefunction', by consciously

observing it, into the condition of being definitely either alive or dead.

7. 'Decoherence' in our everyday world is mainly due to thermal interactions between particles and is incredibly rapid, usually taking less than a femtosecond (i.e. a millionth of a billionth of a second). Although it does not carry quite the same implications as 'collapse of the wavefunction' caused by an observer, the end result is the same for all practical purposes.

8. However, *The Journal of Scientific Exploration* is still going strong and often publishes articles on 'fringe' physics, including proposals about the physics of consciousness. For a detailed but clear and accessible description of why Stapp's particular account of free will is probably unrealistic, see the paper by Bourget listed in the References. Stapp's reply (same journal, pp. 43–9), addresses some of Bourget's objections but skirts the question of whether his (i.e. Stapp's) proposals are physiologically plausible, perhaps because they aren't.

9. Velmans' first paper to attract wide attention was the 1991 one given in the References. In it, he argued that consciousness is a sort of amalgam of events in the brain with events in the outside world. He has continued to publish prolifically on consciousness and related topics, and has written a book for non-specialists.

10. There seems little doubt that genetics does have some influence on our character and behaviour. It would be very surprising if it didn't. However, the influence is often exaggerated in popular accounts. Even identical twins often differ on particular traits. Moreover, although at one time genes were regarded as being permanently available 'read-only' memories that control development and day-to-day functioning, it is now realised that there are all sorts of reciprocal influences, including environmental ones, which may switch genes on and off or influence how vigorously they will be expressed. Although it is not known for sure, it is entirely conceivable that consequences of conscious choice may sometimes be capable of affecting gene expression in the brain. In other words, it is conceivable that choices, or at least their consequences, may sometimes influence genetic activity as well as vice versa.

11. The time between the onset of the electro-negative potential and the experience of volition was around 350 ms, whereas the delay

was often less (typically around 200 ms) in relation to Libet's work on the experience of external information. One can speculate that this difference is due to the fact that external information evoked a sharply defined electrical event in the brain on arrival, whereas the onset of the electronegative potential was not well defined. In general, 300–400 ms seems to be the typical time needed for consciousness to arise. For instance, it is thought that consciousness is associated with the waves occurring at around that time in most work with evoked potentials.

Chapter 5

1. Descriptions of these phenomena can be found in any psychiatric textbook. So-called 'thought insertion' is often a particularly distressing experience – understandably so, since sufferers feel they have no control over what comes into their heads. It has been suggested that auditory hallucinations, another common symptom of schizophrenia, may have a similar origin in that people may misattribute their own sub-vocalisations to an outside source.
2. Capgras syndrome is not confined to schizophrenic patients. People with various forms of organic brain damage, including Alzheimer's disease, can also get it. The opposite condition, when people feel that their kin are inhabiting the bodies of strangers, is called Fregoli syndrome and is rarer than Capgras.
3. See Jablonka and Lamb's book for a wonderful account of how inadequate are purely neo-Darwinian accounts of the development of specific faculties of all sorts.
4. The classical studies of direct electrical stimulation of the brain in conscious patients were undertaken by the Canadian neurosurgeon Wilder Penfield, over a period of around 30 years starting in the 1930s. His patients were mostly epileptics requiring brain surgery. They had to be conscious so that Penfield could ascertain which precise brain areas needed operation. No physical pain was involved, as the brain is not sensitive to pain and local anaesthetics could be given to the scalp. However, going through such a procedure must have required considerable fortitude. Nowadays the brain can be (painlessly) stimulated without any need for operation using TMS (Transcranial Magnetic Stimulation: a pulsed magnetic field). The disadvantage of TMS is that it is not as precise as Penfield's method. The advantage is that it can be applied to volunteers

who are not suffering from any form of brain abnormality. There is also a small 'industry' devoted to studying the consequences of direct electrical stimulation of the brains of primates, which has produced many findings of relevance to humans.

5. Hypnosis, which has nothing to do with sleep despite the name, used to be regarded as a very special state in which the consciousness or mind of the hypnotiser somehow 'took over' that of his (it was usually a 'him') subject. Then the pendulum swung in the 1960s and 1970s towards regarding it as a sort of fake in which subjects simply went along with what was expected of them in the peculiar circumstances of a hypnotic session. The pendulum now seems to be swinging back towards regarding hypnosis as some sort of special state. The swing back is partly due to the fact that 'faking' cannot really explain the physical manifestations in particular that can be elicited, and partly because of the discovery of unusual EEG and fMRI patterns in people who are deeply hypnotised. The literature on the subject is large and confusing. The best single work that I know of is a book by Alan Gauld, which gives the history of the whole subject starting with its origins in Mesmerism.

6. But not *always*. Some subjects will say that they 'felt compelled', or will even deny that they had carried out the suggested action (when in fact they have done so).

Chapter 6

1. There is, in fact, no empirical evidence at all, so far as I know, as to whether infants are consciously aware of 'ownership' of their own actions. At one time, due to the influence of Jean Piaget, many people might have said that they are not. Because of his particular preconceptions and experimental methods, Piaget believed that cognitive abilities in children develop according to a strict schedule. For instance, he thought that babies younger than nine months old cannot conceive of the continuing existence of an object that has disappeared behind a screen. It is now known that he was wrong about this and a number of similar claims. There is a book by Murphy (see References) that discusses these matters in detail. Babies are in fact a lot cannier than was formerly supposed. Since 'responsibility detection' is such an important function when it comes to learning motor and attentional skills, it seems highly likely that infants do have consciousness of ownership of action right from the start.

2. It has been suggested that our concepts are determined by the language that we speak (the Sapir–Whorf hypothesis). However, this is no more than a half or quarter truth. Children certainly often get the concept first and then quickly learn the word to describe it, which is the wrong way round from the Sapir–Whorf point of view. Moreover there is evidence that five-month-old babies have a concept, to do with tightness of fit, which they subsequently *lose* if their native language does not embody the same concept. All the same, there is no doubt that on balance language enormously expands the range of concepts available to us. Without it, we are wholly dependent on our personal experience for our concepts; with it we can both draw on a much wider range of experience and can incorporate concepts developed by other people into our own minds.

3. They were called that then. It stood for Computerised Axial Tomography, now generally shortened to 'CT scanner'. This is an interesting counter-example to the normal 1980s habit of giving as many functions as possible three-word abbreviations – e.g. dustmen (garbage collectors) becoming Waste Disposal Operatives (WDOs), or head cooks resurfacing as Catering Services Managers (CSMs). The name of another type of scanner, developed a bit later, succumbed to a form of political correctness. This was the Nuclear Magnetic Resonance (NMR) scanner, soon changed to Magnetic Resonance Imaging (MRI), for fear that the word 'nuclear' might cause alarm and despondency.

4. As was pointed out in Chapter 4, the question of whether someone could have chosen differently from the way he or she did choose is not relevant to issues to do with whether consciousness may be deterministic in some ultimate physical sense. However, it is relevant to the question of whether conscious 'computation' has some degree of autonomy in relation to unconscious neural computation.

5. The term 'cognitive objects' is deliberately rather vague. These objects certainly include 'memes', a term coined by Richard Dawkins to refer to basic 'units' of cognition like the ideas of building arches, wearing clothes, etc. Dawkins proposed that memes are analogous to genes in that they have a sort of Darwinian competition for resources in our 'mental space'. There are good arguments for the usefulness of this concept, perhaps most clearly expressed in a collection of papers edited by Aunger. Anyone wanting gung-ho accounts of just how marvellous the idea of memes can seem to a

person need look no further than some of Daniel Dennett's books (e.g. *Consciousness Explained* or *Darwin's Dangerous Idea* or *Freedom Evolves*). However, I do not want to use the concept here partly because it has proved impossible to arrive at a definition of memes satisfactory to everyone, while their 'Darwinian' behaviour has sometimes been overplayed. Another reason is that there are other 'cognitive objects' to which I want to refer that are more archetype-like than meme-like. I believe that there are meaningful relationships between these two concepts, but, for the purposes of this book, it is preferable to simply look at how the objects behave rather than try to shoehorn them into any particular theoretical structure. Yet another problem with using the idea of memes in this book is that the 'cognitive objects' to which I refer are at least as much *emotional* as intellectual constructs. As Steven Rose among others has pointed out, cognition is inseparable from emotion. A meme, however, is often considered to be emotion-free. Finally, 'archetype' is itself an unsuitable term as people might assume that it was meant to refer to Jungian archetypes only, which are too restricted and specialised for my purposes.

Chapter 7

1. Shakespeare was historically incorrect here, though he was true to the spirit of the Noble Roman. Brutus was in fact married at least twice and his current wife at the time of the fatal battle (Porcia, daughter of Cato the Younger) almost certainly outlived him as she allegedly committed suicide in a particularly horrible way, by swallowing hot coals, on receiving news of the failure of the Republican cause.
2. C. G. Jung would no doubt have attributed these coincidences to 'synchronicity', the term he used for the sort of meaningful happenings that can occur in people's lives, which must be attributed to chance but nevertheless seem too improbable and apt for this attribution to be true. Of course some of the apparent 'co-incidences' may in fact have been due to early perceptions, by himself and his peers, of Curzon as an avatar of Brutus.

Chapter 8

1. One can speculate that rheumatic chorea and the like may have some *indirect* relationship in that they may have given people the

idea of producing jerky, spasmodic, but nevertheless rhythmic, movements.

2. Hecker was writing at a time when there was a great deal of interest in phenomena of this sort, perhaps sparked by the craze for Mesmerism which had survived Mesmer's enforced retreat from France after his ideas had been discredited by two Royal Commissions.

3. This usage was familiar early in the 19th century; Stewart, in 1827, stated: 'The contagious nature of convulsions, of hysteric disorders, of panics and of all the different kinds of enthusiasm, is commonly referred by medical writers to the principal [sic] of imitation' (*Elements of the Philosophy of the Human Mind*, Murray, London). However, whereas for Hecker 'imitation' seems to have meant something like 'empathy', for Dawkins it refers more to the spread of ideas through word of mouth, education, the media etc.

4. Remarkably, given the gusto with which he wrote, this Lactantius appears to have been the rhetorician and friend of Constantine the Great who was normally against all forms of bloodshed. He was a pacifist and an opponent of capital punishment. Contact with the Millennium has evidently always been liable to distort people's habitual sensibilities.

5. Quotation from Cohn's *The Pursuit of the Millennium*. People sometimes tend to regard the flagellants as by-products of the Black Death. However, their first recorded appearance was around 100 years before the plague.

6. It is interesting that the same seems to be true of the al-Qaeda leadership. They too often seem to be well educated and to come from 'good' families.

7. Gore Vidal provided a characteristically insightful portrait of people like Bockelson in his 1954 novel, *Messiah*. Here is the (fictional) narrator of the book, describing the Messiah:

   He was indifferent, I think, to everyone. He gave one his attention in precise ratio to one's belief in him and the importance of his work. With groups, he was another creature: warm, intoxicating, human, yet transcendent, a part of each man who beheld him, the long desired and pursued whole achieved.

8. Although witches were often burned to death on the continent and in Scotland, they were generally hanged in England when condemned to death at all, which was unusual.

Chapter 10

1. Not to be confused with St Teresa of Avila, a Spanish and far more influential saint who possessed a will of iron, despite (or perhaps because of) her frequent bouts of mystical experience and general 'sensitivity'.
2. William Gladstone, the Victorian Prime Minister, would often pick up prostitutes whom he encountered while out driving in his carriage. He apparently urged them to reform their ways, but was widely suspected of more prurient interests.
3. There is good evidence that literacy skills can influence how the brain develops. For instance, people from illiterate societies tend to use the right side of their brains for language, whereas most of us use the left side. Subtler differences seem to depend on ability to write, not only on ability to read, and perhaps on the availability of printed material in one's society. See Kane's paper in the References for an excellent summary of relevant findings.
4. How times have changed! It seems improbable that any historian nowadays would gratuitously attribute great 'intelligence' to people regarded as proto-medical, whatever his or her views of shamans considered in isolation.
5. But *only* partially. Other factors contributing to the high suicide rate in doctors, for instance, almost certainly include ready access to poisons and knowledge of effective means of ending one's life.
6. Personal information. Following independence, Zambia developed a system in which local personnel called 'medical assistants' were trained at the mental hospital near Lusaka, and sent out to staff small clinics or cottage hospitals in each region. When chronic patients accumulated in the mental hospital, it became necessary to find other accommodation for them or medical assistant training would have become very one-sided. Two abandoned leprosaria were used for this purpose.
7. The idea was probably influenced by the 'leprosarium' or 'asylum' aspect of the Hospital, since it was reserved, at the time, for the

lower social classes. Anyone able to afford it had medical attention at home.

Chapter 11

1. All this was happening in the late 1980s, before the recent high-profile cases of scientific fraud. It was harder then to believe that people might fake results, while whistle blowers were looked on even more unkindly than is the case now.

Chapter 12

1. The Necker cube is a two-dimensional drawing of a cube such that one corner sometimes looks as if it is nearest to you and sometimes looks as if it is furthest away. Quite a lot of ambiguous drawings like this exist, possibly the most intriguing being the one that can look either like a man's face or a seated girl. Some of them vary according to the angle from which one views them, but most simply 'switch' between alternative percepts, purely as a consequence of internal processes of visual perception.
2. This at once suggests that books are likely to retain their importance to cultural continuity for the foreseeable future, because they have a longer shelf life than any of the newer media. For the same reason, they may also be rather better at spreading influences between cultures. Lending libraries may, in the long run, do more than Hollywood blockbusters to extend American culture, for instance.
3. Thanks to the longevity of books, though, Leibnitz is making a bit of a comeback these days. It is too late for him to overtake Newton in creative influence, but it may turn out that he will be remembered for as long.
4. There is a field of study called 'algebraic semantics', which embodies an approach to the mathematical description of meaningful experience in individuals. Although it implicates social networks, it does not explicitly describe them. Nor is it obvious that the approach could be developed so as to model the feedback from individuals to network, needed for any complete description of meaning. Joseph Goguen has provided a fascinating account of some of the issues involved, relating especially to the meaning and experience of music. He also provides a list of relevant references (over 60

of them!), many of which are not readily available from other sources. Goguen's paper can be regarded as an account, which includes suggestions about possible mathematisation, of how short musical 'stories' blend in consciousness to produce perceived music. The complexities involved in even this relatively simple example of how nodes (i.e. us) incorporate stories are formidable, and this is only one side of what would be required for a complete description of the generation of meaning.

5. 'Six degrees of separation' refers to the belief, which is approximately correct in many social networks, that you can generally find a social linkage between any two randomly selected people that has five (or fewer) intermediate steps.

6. When you look in a mirror, you see things (including your own face) as you would do if you were standing behind them. In order to see the same things without a mirror, you would have to stand in front of them and turn through 180° (impossible if it's your own face that you are interested in!). It's this turn which produces the 'normal' left–right ordering. The ordering seen in the mirror, when the turn is not necessary, is the reverse of normal. People often assume that the reversal must be due to some property of the mirror or of one's eyes, but it isn't. It is due to the geometry of the situation. Mirrors would only produce a top-bottom reversal if you had to stand on your head to see whatever you are looking at without the mirror – and you don't.

Chapter 13

1. One particular difficulty may be resolved as new versions of string theory are constantly being developed, some of which may be able to dispense with Newtonian time. Another reason for scepticism is the current inaccessibility of most string theories to experimental tests. This, too, may be less of an obstacle than is often thought, since people have a way of discovering ingenious methods of testing supposedly untestable theories. Interesting discussions of many of the issues involved may be found in Callendar and Huggett's book. See also Sir Roger Penrose's *The Road to Reality* for a detailed account of his reasons for thinking that string theory may represent a blind alley.

2. This realisation has been spurred on by all the work being put into quantum computers and quantum cryptography, both of which

depend on controlling limited entanglements. The main practical difficulty, especially in connection with quantum computation, is to keep entanglement limited for long enough to make use of it. Another spur was the shift from the idea of 'collapse of the wavefunction' to that of 'decoherence'. What happens in decoherence is not that entanglement gets destroyed as it does in wavefunction collapse; rather, a given entanglement gets swamped by a sea of new, random entanglements, usually due to collisions or radiation of thermal origin. A possible, though debatable, consequence of this may be that a measurement of a particle's position, say, is not like seeing where a table is placed. It is more like measuring the temperature of the table, which is a sort of statistical figure deriving from the motions of all of the individual molecular components of the table. A particle's position, when measured, may be a consequence of its entire history of entanglement and that of the measuring instrument; a history which is likely to encompass much of the universe. Although this may seem very strange at first sight, it has long been supposed (since Mach proposed the idea more than a hundred years ago) that a particle's inertia might be a consequence of the existence of the whole of the rest of the universe, which is a not dissimilar concept.

3. In something of a *tour de force*, Roy Frieden was able to show that many of the most fundamental equations of physics, including Schrödinger's, can be derived from 'Fisher information', which is a quantified measure of our degree of ignorance of some phenomenon. What these equations do is to minimise our ignorance of the situations in which they apply. Interestingly, the quantum equations are better at doing this than most classical equations. There is a relationship between Fisher information and the more familiar Shannon information (i.e. our customary bits and bytes), but its precise nature is not clear – at least, it's not clear to me. Frieden's work provides a neat demonstration that the wavefunction, and indeed physics in general, is to do with information, even though the type of information most useful from his point of view turned out to be a little different from the usual concept.

4. Wheeler is among the greatest of modern physicists, as well as the most long-lived (born 1911). Unlike many very creative people, he was also a wonderful mentor and a thoroughly decent man. Richard Feynman, the most famous of his students, was one among many who went on to do great things themselves. Wheeler wrote an

autobiography, *Geons, Black Holes and Quantum Foam*, which is well worth reading.

5. I don't want to provide the details here as they are not relevant, and have often been described elsewhere. In fact the apparatus has been greatly refined recently. Whereas the variant described by Penrose in *Shadows of the Mind* will only sometimes tell you that a bomb is live without causing an explosion, more elaborate versions give a chance of as little as 10% of exploding a live bomb. Of course, the version involving a 'bomb' is a thought experiment only, but equivalent setups have been built and tested.
6. David Bohm tried very hard to persuade people that a picture like this might correspond to what's really out there, with his notion that particles do have definite positions, momenta etc., but are subject to the guidance of an immaterial 'pilot wave'. The picture works, but seems so contrived that it has not appealed to many physicists.
7. By Bekenstein, another student of John Wheeler's. Lee Smolin has provided a most lucid and readable account of all this: *Three Roads to Quantum Gravity*.
8. There are persistent claims that people can access information about the physical world while 'detached' from their bodies. One woman, for instance, allegedly 'saw' a shoe on an outside window ledge, while floating through another window. The existence of the shoe was later confirmed by a nurse who went to look for it, so it was claimed. Others have reported details of surgical procedures and the like that they could only have seen from a vantage point outside their bodies – or so it is said. Several attempts have been made to test such stories, for instance by concealing pictures above the ceilings of intensive care units in the hope that some souls would spot them on their way up. So far as I know, no convincing confirmation has been derived from tests like these. What evidence there is remains strictly anecdotal.
9. See Strassman's book, *DMT, the Spirit Molecule.* This is a thoughtful and sensible account of the properties of, and clinical research on, DMT, a hallucinogen that is produced by the brain itself. Although it is usually destroyed too quickly to have much effect, Strassman suggests that larger amounts may accumulate in some circumstances and cause a number of phenomena, including NDEs.

10. Shanon is a professor of psychology, based in Jerusalem, who first became interested in ayahuasca while on a visit to Ecuador in 1983. His personal experience is based on taking the drug in a variety of both Amazonian and 'Western' settings. Cynics may wonder whether he might have been damaged by the drug. There is no evidence whatsoever from the book that this might be the case. It is of a high scientific standard, as well as being a most thoughtful and fascinating document. He himself, as he describes in the book, found his experiences to be spiritually enriching. It is clear from his descriptions of the physical side-effects alone, though, that one needs a lot of courage, fortitude and physical stamina in order to benefit.

11. Jeremy Narby's very readable and thought-provoking book *The Cosmic Serpent* contains interesting observations on the phenomenology of shamanic visions, and imaginative (but in many ways far-fetched) speculations about their relationship to the nature of DNA.

12. The loss of insight that can occur in 'toxic' psychosis is exemplified by a former patient of mine, who had been indulging in too much Guinness for too many years. He was a successful building contractor living in a city, but had been raised in the countryside. One morning, he opened his front door and, instead of the familiar street, saw before him a field of wheat that needed cutting. He started trying to get passing commuters to help him with this. Shanon notes that ayahuasca drinkers often retain insight into the hallucinatory nature of their experiences, but the 'sense of reality' may itself be illusory and not subject to this insight.

13. Prior to Einstein's explanation of the discrepancy on the basis of general relativity, astronomers often thought that it might be due to an undiscovered planet closer to the sun than Mercury, and indeed put some effort into searching for the object. They gave it a name – Vulcan – but of course it did not exist.

14. Karl Pribram, one of the greatest of contemporary neurophysiologists, believes that both matter and mind are different aspects of what he terms 'flux', a concept which appears to be a generalised version of the notion of a 'story' developed here. He has a nice anecdote about how he, tongue in cheek, asked an eminent quantum physicist (Eugene Wigner) whether 'quantum physics is really psychology?' Instead of the expected gruff reply, Wigner gave

a happy smile and said 'Yes, yes, that's exactly correct.' In so far as quantum physics describes our best current understanding of the nature of matter, while 'free will' is an aspect of psychology, this reply can be taken to suggest that the two (i.e. matter and 'free will') may not be so totally different as is commonly supposed. Niels Bohr, the 'George Washington' of quantum physics, seems to have shared this view. According to Meyer-Abich, he said that the results of quantum physics 'reminded' him of psychology. When two things share the same essential nature, causative relationships between them may appear to flow in either direction, according to the contexts in which they are viewed and the type of interaction they are undergoing. Anyone sufficiently interested in current speculations about relationships between mind, physics, information and meaning can find a collection of essays by several of the main 'players' in this field in *Brain and Being* (eds. Gordon Globus, Karl Pribram and Giuseppe Vitiello), John Benjamins Publishing Co., 2004. These are a bit more readable than most discussions of these topics.

Epilogue

1. Both Margaret Thatcher and Henry Ford have been unlucky in that their famous remarks are so often quoted out of context. In context they were not nearly so silly. On the other hand, in Thatcher's case at least, those of us who remember her assertion that 'the National Health Service is safe in our hands' may uncharitably be tempted to think that she deserves all she gets!

2. Of course, the most basic aspects of our lives derive from our biology and this too could be regarded as an expression of story lines, though in this case biochemical and evolutionary ones. If one does generalise the notion of a 'story' in this way, then the basic aspects of our lives have far longer histories than the aspects deriving from culture.

REFERENCES

Chapter 1

Alexander, D. (2001) *Rebuilding the Matrix: Science and Faith in the 21st Century*. Lion Books, Oxford.

Freeman, A. (2000) Responsibility without choice: a first person approach. *Journal of Consciousness Studies*, **7**(10), 61–7.

Journal of Consciousness Studies website: http://www.imprint.co.uk/jcs/

Chapter 2

Baars, B. (1997) *In the Theatre of Consciousness: the Workspace of the Mind*. Oxford University Press, Oxford.

Baars, B. J. and McGovern, K. (1996) Cognitive views of consciousness. In Velmans, M. (ed.) *The Science of Consciousness*. Routledge, London.

Bailey, L. and Yates, J. (eds.) (1996) *The Near Death Experience: a Reader*. Routledge, London.

Cossart, R., Aronov, D. and Yuste, R. (2003) Attractor dynamics of network UP states in the neocortex. *Nature*, **423**, 283–88.

Churchland, P. M. (1995) *The Engine of Reason, the Seat of the Soul*. MIT Press, Cambridge MA.

Crick, F. and Koch, C. (1992) The problem of consciousness. *Scientific American*, September, pp. 111–17.

Dennett, D. C. (2003) Who's on first? Heterophenomenology explained. *Journal of Consciousness Studies*, **10**(9–10), 19–30.

Eccles, J. and Popper, K. (1977) *The Self and its Brain*. Springer International, New York.

Edelman, G. M. (1989) *The Remembered Present: a Biological Theory of Consciousness*. Basic Books, New York.

Elitzur, A. C. (1989) Consciousness and the incompleteness of the physical explanation of behaviour. *Journal of Mind and Behaviour*, **10**(1), 1–20.

Freeman, W. (1999) *How Brains Make up their Minds*. Weidenfeld & Nicholson, London.

Hameroff, S. and Penrose, R. (1996) Conscious events as orchestrated space–time selections. *Journal of Consciousness Studies*. **3**(1), 36–53.

Harth, E. (1993) *The Creative Loop: How the Brain Makes a Mind.* Penguin, London.

Hofstadter, D. R. (1979) *Gödel, Escher, Bach: an Eternal Golden Braid.* Penguin, London.

Jibu, M. and Yasue, K. (1995) *Quantum Brain Dynamics and Consciousness.* John Benjamins Publishing, Amsterdam and Philadelphia.

Kim, J. (2000) *Mind in a Physical World.* MIT Press, Cambridge MA.

Koch, C. (1999) *Biophysics of Computation: Information Processing in Single Neurons.* Oxford University Press, Oxford.

Lamme, V. A. and Spekreijse, H. (1998) Neural synchrony does not represent texture segregation. *Nature*, **396**, 362–6.

MacCormac, E. and Stamenov, M. (eds.) (1996) *Fractals of Brain, Fractals of Mind.* John Benjamins Publishing, Philadelphia and Amsterdam.

Marshall, I. N. (1989) Consciousness and Bose–Einstein condensates. *New Ideas in Psychology*, **7**(1), 73–83.

Meier, C. A. (ed.) (2001) *Atom and Archetype; the Jung/Pauli letters, 1932–1958.* London, Routledge.

Nagel, T. (1974) What is it like to be a bat? *Philosophical Review*, **83**, 435–50.

Savage-Rumbaugh, E. S. and Lewin, R. (1994). *Kanzi; the Ape At the Brink of the Human Mind.* John Wiley, New York.

Savage-Rumbaugh, S., Fields, W. and Taglialatela, J. (2001) Language, speech, tools and writing. *Journal of Consciousness Studies*, **8**(5–7), 273–92.

Seager, W. (1995) Consciousness, information and pan-psychism. *Journal of Consciousness Studies*, **2**(3), 272–88.

Skarda, C. A. and Freeman, W. J. (1987) How brains make chaos in order to make sense of the world. *Behavioural and Brain Sciences*, **10**, 161–195.

Skrbina, D. (2003) Panpsychism in Western Philosophy. *Journal of Consciousness Studies*, **10**(3), 4–46.

Strawson, G. (1994) *Mental Reality.* MIT Press, Cambridge, MA.

Thiele, A. and Stoner, G. (2003) Neural synchrony does not correlate with motion coherence in cortical area MT. *Nature*, **421**, 366–70.

Varela, F., Thompson, E. and Rosch, E. (1993) *The Embodied Mind: Cognitive Science and Human Experience.* MIT Press, Cambridge MA.

Chapter 3

Bickle, J. (2003) *Philosophy and Neuroscience: a Ruthlessly Reductive Account*. Kluwer Academic, New York.

Eichenbaum, H. (1999) The topography of memory. *Nature*, **402**, 597–99.

Eichenbaum, H. and Cohen, N. J. (2001) *From Conditioning to Conscious Recollection: Memory Systems of the Brain*. Oxford University Press, Oxford.

Hebb, D. O. (1949) *The Organisation of Behaviour.* Wiley, New York.

Ito, M. (2000) Internal model visualised. *Nature*, **403**, 153–4.

Malenka, R. and Nicoll, R. (1997) Never fear, LTP is here. *Nature*, **390**, 552–3.

Nader, K., Schafe, G. E. and Le Doux, J. (2000) Fear memories require protein synthesis in the amygdala after retrieval. *Nature*, **406**, 722–6.

Nicolis, J. and Tsuda, I. (1989) Chaotic dynamics of information processing: the magic number seven plus-minus two revisited. *Bulletin of Mathematical Biology*, **47**(3), 343–65.

O'Craven, K., Downing, P. and Kanwisher, M. (1999) fMRI evidence for objects as the units of attentional selection. *Nature*, **401**, 584–87.

Pribram, K. (2000) Neuropsychological investigations. In van Loocke, P. (ed.) *The Physical Nature of Consciousness*. John Benjamins Publishing, Amsterdam and Philadelphia.

Todd, J. and Marois, R. (2004) Capacity limit of visual short-term memory in human posterior parietal cortex. *Nature*, **428**, 751–4.

Tseng, C.-H., Gobell, J. and Sperling, G. (2004) Long-lasting sensitisation to a given colour after visual search. *Nature*, **428**, 657–60.

Watt, D. and Pincus, D. (2004) The neural substrates of consciousness. In Panksepp, J. (ed.) *Textbook of Biological Psychiatry.* John Wiley, New Jersey.

Chapter 4

Bourget, D. (2004) Quantum leaps in philosophy of mind. *Journal of Consciousness Studies*, **11**(12), 17–42.

Fenwick, P. (1993) Brain, mind and behaviour; some medico-legal aspects. *British Journal of Psychiatry*, **163**, 565–73.

Gomes, G. (1999) Volition and the readiness potential. *Journal of Consciousness Studies*, **6**(8–9), 59–76.

Hodgson, D. (1999) Hume's mistake. *Journal of Consciousness Studies*, **6**(8–9), 201–24.

Hodgson, D. (2002) Three tricks of consciousness. *Journal of Consciousness Studies*, **9**(12), 65–88.

Hofstadter, D. (1986) *Metamagical Themas*. Penguin, London.

Libet, B. (1989) Conscious subjective experience vs unconscious mental functions: a theory of the cerebral processes involved. In Cotterill, R. (ed.) *Models of Brain Function*. Cambridge University Press, Cambridge.

Libet, B. (1994) A testable field theory of mind–brain interaction. *Journal of Consciousness Studies*, **1**(1), 119–26.

Libet, B. (1996) Neural processes in the production of conscious experience. In Velmans, M. (ed.) *The Science of Consciousness*. Routledge, London.

Libet, B. (1999) Do we have free will? *Journal of Consciousness Studies*, **6**(8–9), 47–57.

Libet, B., Gleason, C., Wright, E. and Pearl, D. (1983) Time of conscious intention to act in relation to onset of cerebral activity. *Brain*, **106**, 623–42.

Pockett, S. (2002) Difficulties with the electromagnetic field theory of consciousness. *Journal of Consciousness Studies*, **9**(4), 51–6.

Pockett, S. (2004) Does consciousness cause behaviour? *Journal of Consciousness Studies*, **11**(2), 23–40.

Radin, D. (1997). *The Conscious Universe*. HarperEdge, San Francisco.

Spence, S. and Frith, C. (1999) Towards a functional anatomy of volition. *Journal of Consciousness Studies*, **6**(8–9), 11–29.

Stapp, H. (1993) *Mind, Matter and Quantum Mechanics*. Springer-Verlag, New York.

Stapp, H. P. (1999) Attention, intention and will in quantum physics. *Journal of Consciousness Studies*, **6**(8–9), 143–64.

Trevena, J. and Miller, J. (2002) Cortical movement preparation before and after a conscious decision to move. *Consciousness and Cognition*, **11**, 162–90.

Velmans, M. (1991) Is human information processing conscious? *Behavioral and Brain Sciences*, **14**(4), 651–69.

Velmans, M. (2000) *Understanding Consciousness*. Routledge, London.

Velmans, M. (2002) How could conscious experiences affect brains? *Journal of Consciousness Studies*, **9**(11), 3–29.

Wall, P. (1996) The placebo effect. In Velmans, M. (ed.) *The Science of Consciousness*. Routledge, London.

Wolfram, S. (2002) *A New Kind of Science*. Wolfram Media.

Chapter 5

Gauld, A. (1992) *A History of Hypnotism*. Cambridge University Press, Cambridge.

Hohwy, J. and Frith, C. (2004) Can neuroscience explain consciousness? *Journal of Consciousness Studies*, **11**(7–8), 180–98.

Jablonka, E. and Lamb, M. J. (2005). *Evolution in Four Dimensions*. MIT Press, Cambridge MA.

Wegner, D. (2002) *The Illusion of Conscious Will*. MIT Press, Cambridge MA.

Wegner, D. and Wheatley, T. (1999) Apparent mental causation: sources of the experience of will. *American Psychologist*, **54**, 480–91.

Chapter 6

Aunger, R. (ed.) (2000) *Darwinizing Culture: the status of Memetics as a Science*. Oxford University Press, Oxford.

Bloom, P. (2004) Children think before they speak. *Nature*, **430**, 410–11.

Merriman, W., Schuster, J. and Hager, L. (1991) Are names ever mapped onto pre-existing concepts? *Journal of Experimental Psychology: General*, **120**, 288–300.

Murphy, G. (2002) *The Big Book of Concepts*. MIT Press, Cambridge MA.

Nunn, C. (1998) Archetypes and memes: their structure, relationships and behaviour. *Journal of Consciousness Studies*, **5**(3), 344–54.

Rose, S. (2005) *The 21st Century Brain: Explaining, Mending and Manipulating the Mind*. Jonathan Cape, London.

Chapter 7

Grant, M. (1978) *History of Rome*. Faber & Faber, London.

King, P. (1986) *The Viceroy's Fall: How Kitchener Destroyed Curzon*. Sidgwick & Jackson, London.

Rose, K. (1969) *Superior Person: a Portrait of Curzon and his Circle in Late Victorian England*. Weidenfeld & Nicolson, London.

Chapter 8

Ball, P. (2003) *Critical Mass: How One Thing Leads to Another*. William Heinemann, London, p. 395.

Bartholomew, R. E. (1994) Tarantism, dancing mania and demonopathy: the anthro-political aspects of 'mass psychogenic illness'. *Psychological Medicine*, **24**, 281–306.

Cohn, N. (1975) *Europe's Inner Demons: an Enquiry Inspired by the Great Witch-Hunt.* Chatto Heinemann/Sussex University Press, London.

Hecker, J. F. C. (1837) *The Dancing Mania of the Middle Ages* (1885 edn). Humboldt Publishing Co., New York.

MacDonald, M. (1981) *Mystical Bedlam: Anxiety and Healing in Seventeenth Century England.* Cambridge University Press, Cambridge.

Mcleod, L. (1865) *Madagascar and its People.* Longman and Green, London.

Mumford, L. (1965) Utopia, the city and the machine. In *Interpretations and Forecasts 1922–1972.* Harcourt Brace Jovanovich, New York.

Norman, H. E. (1945) Mass hysteria in Japan. *Far Eastern Survey*, **14**(6), 65–70.

Runciman, S. (1954) *A History of the Crusades* (3 volumes). Penguin, London.

Chapter 9

Blackmore, R. (1725) *A Treatise of the Spleen and Vapours: or Hypochondriacal and Hysterical Affections.* Pemberton, London.

Boyle, R. (1691) *Experimenta & Observationes Physicae: wherein are briefly treated of several subjects relating to natural philosophy in an experimental way. To which is added a small collection of strange reports.* Taylor & Wyat, London.

Bynum, W. F. (1985) the nervous patient in 18th and 19th century Britain. In Bynum, W. F., Porter, R. and Shepherd, M. (eds.) *The Anatomy of Madness*, Vol. 1. Tavistock Publications, London.

Cheyne, G. (1733) *The English Malady; or, a Treatise of Nervous Disorders of All Kinds, as Spleen, Lowness of Spirits, Hypochondriacal and Hysterical Distempers &c.* Strahan & Leake, London.

Da Costa, J. M. (1871) On irritable heart; a clinical study of a form of functional cardiac disorder and its consequences. *Journal of the American Medical Society*, **61**, 17.

Delamothe, T. (1994) Look at ME. *British Medical Journal*, **308**, 798.

Gallagher, A. M. *et al.* (2004) Incidence of fatigue syndromes and diagnoses presenting in UK primary care from 1990 to 2001. *Journal of the Royal Society of Medicine,* **97**, December, 571–5.

Graunt, J. (1662) *Natural and Political Observations Mentioned in a Following Index, and Made upon the Bills of Mortality, 1662*. Martin *et al.*, London.

Kendell, R. E. (1993) Chronic Fatigue Syndrome. *Lancet*, **341**, 1137.

Macintyre, A. (1989) *ME. Post-Viral Fatigue Syndrome: How to Live With It.* Unwin Paperbacks, London.

Merskey, H. (1991) Shell-shock. In Berrios, G. E. and Freeman, H. (eds.) *150 Years of British Psychiatry*. Gaskell, London.

Pawlikowska, T., Chalder, T., Hirsch, S. R., Wallace, P., Wright, D. J. M. and Wessely, S. C. (1994) Population based study of fatigue and psychological distress. *British Medical Journal*, **308**, 763–6.

Pierce, R. (1697) *Bath Memoirs: or, Observations in Three and Forty Years of Practise, At the Bath, What Cures Have There Been Wrought.* Hammond, Bristol.

Ramsay, A. M. (1988) *Myalgic Encephalomyelitis and Postviral Fatigue States*, 2nd edn. Gower Medical, London.

Riccio, M., Thompson, C. and Wilson, B. (1992) Neuropsychological and psychiatric abnormalities in myalgic encephalomyelitis: a preliminary report. *British Journal of Clinical Psychology*, **31**, 111–120.

Wessely, S. (1990) Old wine in new bottles: neurasthenia and ME. *Psychological Medicine*, **20**, 35–53.

Wilson, A., Hickie, I., Lloyd, A., Hadzi-Pavlovic, D., Boughton, C., Dwyer, J. and Wakefield, D. (1994) Longitudinal study of outcome of chronic fatigue syndrome. *British Medical Journal*, **308**, 756–9.

Wookey, C. (1986) *Myalgic Encephalomyelitis: Post-Viral Fatigue Syndrome and How to Cope With It.* Chapman & Hall, London.

Chapter 10

Bynum, W. F., Porter, R. and Shepherd, M. (1988). Introduction. In *The Anatomy of Madness, Vol. 3: The Asylum and its Psychiatry.* Routledge, London.

Camp, J. (1978) *The Healer's Art: the Doctor Through History.* Frederick Muller Ltd, London.

Caraman, P. (1990) *Ignatius Loyola.* Collins, London.

De la Bedoyere, M. (1960). *François de Sales*. Collins, London.

Hollings, M. (1981) *Thérèse of Lisieux.* Collins, London.

Jones, K. (1972) *A History of the Mental Health Services*. Routledge & Kegan Paul, London.

Kane, J. (2004) Poetry as right-hemispheric language. *Journal of Consciousness Studies*, **11**(5–6), 21–59.

Mercier, C. A. (1915) *Leper Houses and Mediaeval Hospitals*. London.
Mockler, A. (1976) *Francis of Assisi: the Wandering Years*. Phaidon, Oxford.
Rahner, H. (1960) *Saint Ignatius: Letters to Women*. Freiburg, Edinburgh and London.
Rhodes, P. (1985) *An Outline History of Medicine*. Butterworth, London.
Rodger, N. A. M. (2004) *The Command of the Ocean*. Allen Lane, London, p. 399.
Rosen, G. (1959) *A History of Public Health*, 1993 edn. Johns Hopkins University Press, Baltimore.
Rubin, S. (1974) *Medieval English Medicine*. David & Charles, Newton Abbott.
Spiro, H. M. (1993) Empathy: an introduction. In Spiro, H., McCrea Curnen, M., Peschel, E. and St. James, D. (eds.) *Empathy and the Practice of Medicine*. Yale University Press, New Haven CT.
Stevenson, C. (1988) Madness and the picturesque in the Kingdom of Denmark. In Bynum, W. F., Porter, R. and Shepherd, M. (eds.) *The Anatomy of Madness, Vol. 3: The Asylum and its Psychiatry*. Routledge, London.
Williams, G. (1975). *The Age of Agony*. Constable, London.

Chapter 11

Damasio, A. R. (1999) *The Feeling of What Happens: Body and Emotion in the Making of Consciousness*. Harcourt, London.

Chapter 12

Dalai Lama (2003) *Stages of Meditation*. Snow Lion Publications, New York.
Devlin, K. (1991) *Logic and Information*. Cambridge University Press, Cambridge.
Devlin, K. (1997) *Goodbye Descartes: the End of Logic and the Search for a New Cosmology of the Mind*. Wiley, New York.
Goguen, J. (2004) Musical qualia, context, time and emotion. *Journal of Consciousness Studies*, **11**(3–4), 117–47.
Gray, R. M. (1996) *Archetypal Explorations*. Routledge, London.
Hutchins, E. (1995) *Cognition in the Wild*. MIT Press, Cambridge MA.
Pagel, M. and Mace, R. (2004) The cultural wealth of nations. *Nature*, **428**, 275–8.

Chapter 13

Bailey, L. and Yates, J. (eds.) (1996) *The Near Death Experience: a Reader.* Routledge, London.

Blackmore, S. (1993) *Dying to Live: Science and the Near Death Experience*. Grafton, London.

Bohm, D. and Hiley, B. (1993)*The Undivided Universe*. Routledge, London.

Callendar, C. and Huggett, N. (2001). *Physics Meets Philosophy at the Planck Scale*. Cambridge University Press, Cambridge.

Fenwick, P. and Fenwick, E. (1995). *The Truth in the Light: an Investigation of over 300 Near Death Experiences*. Headline, London.

Fox, M. (2003) *Religion, Spirituality and the Near Death Experience.* Routledge, London.

Frieden, R. (1998), *Physics from Fisher Information*. Cambridge University Press, Cambridge.

Hackermuller, L., Hornberger, K., Brezger, B., Zeilinger, A. and Arndt, M. (2004) Decoherence of matter waves by thermal emission of radiation. *Nature*, **427**, 711–14.

Hanke, D. (2003) Teleology: the explanation that bedevils biology. In Cornwell, J. (ed.) *Explanations: Styles of Explanation in Science.* Oxford University Press, Oxford.

Meyer-Abich, K. (2004) Bohr's complementarity and Goldstein's holism. *Mind and Matter,* **2**(2), 92–103.

Moody, R. (1975) *Life After Life*. Mockingbird, Atlanta.

Narby, J. (1999) *The Cosmic Serpent*, 2003 edn. Phoenix, London, p. 148.

Penrose, R. (2004) *The Road to Reality*. Jonathan Cape, London.

Pribram, K. (2004) Consciousness reassessed. *Mind and Matter,* **2**(1), 7–35.

Shanon, B. (2002) *The Antipodes of the Mind: Charting the Phenomenology of the Ayahuasca Experience*. Oxford University Press, Oxford.

Smolin, L. (2000). *Three Roads to Quantum Gravity.* Weidenfeld & Nicolson, London.

Strassman, R. (2001). *DMT, the Spirit Molecule*. Park Street Press, Rochester.

Wheeler, J. (1998) *Geons, Black Holes & Quantum Foam.* W. W. Norton & Co., New York.

White, R. (1997) Exceptional human experiences and the experiential paradigm. In Tart, C. (ed.) *Body, Mind, Spirit*. Hampton Roads, Charlotteville.

INDEX